41.99

# The LAN Manager's
# Internet Connectivity Guide

# THE MCGRAW-HILL SERIES ON COMPUTER COMMUNICATIONS (SELECTED TITLES)

# The LAN Manager's Internet Connectivity Guide

Dr. Sidnie Feit

**McGraw-Hill**
New York • San Francisco • Washington, D.C.
Auckland • Bogotá • Caracas • Lisbon • London • Madrid
Mexico City • Milan • Montreal • New Delhi • San Juan
Singapore • Sydney • Tokyo • Toronto

**Library of Congress Cataloging-in-Publication Data**

Feit, Sidnie.
    The LAN manager's Internet connectivity guide / Sidnie Feit.
        p.     cm.—(McGraw-Hill series on computer communications)
    Includes index.
    ISBN 0-07-061622-1
    1. Local area networks (Computer networks)  2. Internet (Computer
network)  I. Title.  II. Series.
    TK5105.7.F44   1997
    004.67'8—dc21
                                                         97–24715
                                                           CIP

# McGraw-Hill

*A Division of The **McGraw·Hill** Companies*

1 2 3 4 5 6 7 8 9 0  FGR/FGR  9 0 2 1 0 9 8 7

ISBN 0-07-061622-1

*The sponsoring editor for this book was Steve Elliot and the production supervisor was Sherri Souffrance. It was set in Vendome by Graphic World, Inc.*

*Printed and bound by Fairfield Graphics.*

*McGraw-Hill books are available at special quantity discounts to use as premiums and sales promotions, or for use in corporate training programs. For more information, please write to Director of Special Sales, McGraw-Hill, 11 West 19th Street, New York, NY 10011. Or contact your local bookstore.*

 This book is printed on recycled, acid-free paper containing a minimum of 50% recycled, de-inked fiber.

# CONTENTS

Contents

Contents

Contents

# Contents

# PREFACE

As the title states, this book is intended to be a guide for both managers and technical staff who are responsible for connecting sites to the Internet. For technical readers, we have tried to focus on the specific protocol and implementation internals that they will need to know in order to get a site connected, keep performance at a high level, and maintain site security.

The protocol tutorials are structured so that they can be read by a beginner. However, the discussion quickly zooms in on the details that relate to server and router configuration or deal with performance and security. Even if you have quite a lot of experience with the TCP/IP protocol family, we hope that you can benefit from a walk-through that explains performance and security implications at each step along the way.

Although some of the book's material is fairly technical, we hope that large parts of the book will be readable and valuable to non-technical managers who will:

- Learn something about the benefits that an Internet connection can provide.
- Get pointers to ideas for new services and products.
- Understand guidelines that will help to make their organization's Web site perform up to its potential and serve its clients well.
- Gain an understanding of security problems and solutions.

# Organization of the Book

Chapter 1 explains how the Internet came into existence and how it is put together today. Names, addresses, routing, and Internet client/server concepts are introduced.

Chapter 2 introduces World Wide Web concepts and describes standard types of business sites. Real world examples are used to illustrate the innovative ways in which a Web site can be used to improve service, reduce costs, or create business opportunities. The chapter also presents techniques that an organization can use to help potential clients to find their site. The benefits of search tools are described, as well as methods of getting good visibility for your site at popular search centers. The chapter ends with a survey of common Web design errors that cripple performance and often lock out potential visitors.

Prior to connecting to the Internet, it is important for key personnel to get connected via dial-up accounts. Many people already have such accounts. Chapter 3 briefly describes the type of access that will be most useful and explains how you can track down and evaluate an *Internet Service Provider* (ISP) that offers an appropriate dial-up service. The chapter also explains some software installation and configuration issues.

In Chapter 4, we introduce the first issue that you will have to deal with before connecting your site to the Internet—name registration. Detailed information is provided that will help you to maintain control of your registration and update your information securely. There are times when you will need to locate the administrators responsible for other sites in order to solve a performance or security problem. You will find out how to do this quickly.

You need to choose on-site equipment that is the most appropriate way to connect your site to the Internet. Chapter 5 will describe the merits of filtering routers, network address translators, and several types of firewalls. You also need to select a transmission technology that is consistent with your performance requirements and your budget. Chapter 5 describes and compares the choices of transmission technologies that are offered by Internet Service Providers. Chapter 6 provides criteria that can help you to select the best ISP for your site.

Chapter 7 presents Internet address formats and explains how sites can avoid the address shortage crunch by using *Private Enterprise Addresses*. The chapter presents a detailed explanation of how you can function well with very small Internet address allocations and offers strategies for minimizing renumbering problems.

Technical information about TCP/IP internals is provided in Chapter 8. The focus is on information that is needed in order to understand performance and security issues.

Chapter 9 is devoted to the router that connects a site to the Internet. This chapter fills in the many technical details needed in order to understand router configuration parameters.

The fact that the Internet evolved as a research project within a mutually trusting community led to one omission—a lack of security mechanisms. This omission is being remedied today, and we will describe security threats and solution products in Chapter 10. Electronic signatures and certification, which are essential for electronic commerce, also will be explained.

Chapter 11 describes how Internet mail works, including the essential role of the Internet's directory service, the Domain Name System. Important features of current products are presented, including security functionality.

Every Internet site is responsible for supporting its part of the Domain Name System. Chapter 12 describes how the whole structure works and contains information that is important to administrators and help desk personnel. Chapter 12 contains detailed examples that show how DNS servers are set up on both NT and Unix platforms.

Chapter 13 is devoted to World Wide Web technology and performance and describes a number of ways in which Web clients and servers are being enhanced and extended. The chapter also presents the mechanisms that underlie forms, secure forms, and Common Gateway Interface programs. Chapter 14 provides an overview of two Web server products: Microsoft Internet Information Server and the Unix Apache server.

# ACKNOWLEDGMENTS

I wish to thank H. Morrow Long for sharing his boundless knowledge of operating systems and networks, and teaching me what I know about security. Graham Yarbrough answered many questions along the way with patience and good humor. Both Gary Kessler and Graham Yarbrough read through the manuscript with great care and made many helpful suggestions.

My thanks to Rich Klener and Cisco Systems for the loan of a Cisco 2503 router. Bob Williams at Netmanage provided the Chameleon TCP/IP client and server software for Windows 3.1, Windows 95, and Windows NT that was the basis for many tests and experiments.

Scott Schmaling of Southern New England Telephone, which provides Internet connectivity services, answered many questions about telephone line interfaces. Jerimy Black at Internet Service Provider VNET described several configurations for client sites. Both offered insights into Service Provider support issues. Jim Safran of VNET had great patience in discussing Service Provider business issues, and supplied the sample contract that appears in this book.

I also wish to thank Data-Tech Institute, which provided an NT server system and some of the software used in demonstrations in this book.

# INTRODUCTION

An Internet connection offers many opportunities for companies to save money, improve customer assistance, and introduce new products and services. However, browsing the Internet today, it is unfortunately obvious that there are many organizations whose management has not taken the time to become familiar with the Internet environment and formulate business goals before connecting their site or setting up a Web server.

An organization should connect to the Internet to promote some identified business or organizational objectives. The benefits of plugging into the Internet will be greatest when the goals are explicit and are well understood by those responsible for connecting the site or setting up its World Wide Web server.

Even when there is a well-understood business purpose, the technical side of the implementation often is carried out poorly. Frequently it is clear that management personnel and Web designers have never visited their own Web site via a dial-up Internet connection and experienced the frustration and long waits that their customers must endure. Or perhaps a manager has tried out such a connection and was told that "the Internet is slow today," when in fact the problem is caused by a bundle of design flaws.

## Dialogs

All of the major points in this book are illustrated by graphical figures or computer dialogs. The dialogs were generated by interactions between systems running Windows 3.1, Windows 95, Windows NT Server, and several flavors of Unix. Both text and graphical user interfaces are shown. The Windows graphical user interface displays show built-in screens for Windows 95 and NT Server, Netmanage Chameleon applications, Netscape Inc.'s *Netscape Navigator*, Microsoft's *Internet Explorer*, and Ashmount Research Ltd.'s freely available *NSLookup for Windows*.

## Internet References

The book includes many references to specific Internet sites and online documents that provide useful information. However, the Internet is a volatile environment. Names and locations tend to shift over time. We have included information about Internet searching that should help you to track down reference material that has moved.

## Text Conventions

In computer dialogs, commands that the user enters will appear in boldface. For example:

```
> nslookup
Default Server: DEPT-GW.CS.YALE.EDU
Address: 128.36.0.36
```

Within the text, italics will be used for several items, including:

- Command names, such as *traceroute* and *nslookup*.
- Computer names, such as *www.mcgraw-hill.com*.
- File names, such as *index.html*.
- World Wide Web Uniform Resource Locators, such as:

  *http://rs.internic.net/rs-internic.html*

Menu titles will be written in Helvetica font: for example, File. When you need to go through a chain of menus to reach an item, the menus will be separated by forward slashes. For example, to configure network properties for Microsoft Windows 95, you would traverse:

Start/Settings/Control Panel/Network

## Bits and Bytes

A bit is the smallest amount of information that can be stored or transmitted, and has a value of 0 or 1. As used in this book, a byte is 8 bits. Data communications speeds often are measured in bits per second. Computer memory and disk space are measured in bytes.

## Communications Kilo and Mega

Communications people measure data transmission rates in bits per second, kilobits per second, and megabits per second. To a communications person, "kilo" means "thousand." Thus, a rate of 64 kilobits per second (also written 64 Kbps) means 64,000 bits per second. Similarly, to a communications person, "mega" means "million," and a rate of 2 megabits per second (also written 2 Mbps) means 2 million bits per second.

## Data Kilo and Mega

Computer data people measure data quantities in bytes, kilobytes, and megabytes. However, data measurements increase by powers of 2. To a data person, "kilo" means $2^{10} = 1024$. Thus a file that contains 28 kilobytes (also written KB) holds 28,672 bytes. Similarly, to a data person, "mega" means $2^{20} = 1,048,576$. Thus, a file that contains 10 megabytes (also written MB) holds 10,485,760 bytes.

## Networking Terms

Several common network performance terms will be used in this book:

- *Response time* is the performance measurement that matters most to users. It is the time that it takes for them to get what they ask for. End-user response time may be slow because of delay on the network, delay at the server that has to carry out the user's request, or a combination of both. Occasionally, we will isolate and examine the network part of the response time.

- *Bandwidth* originally meant the range of electrical frequencies that a communications circuit could utilize. Today, bandwidth is typically a measurement of the number of bits that can be sent through a given communications circuit in a given unit of time.

- *Throughput* also measures the rate of transmission of information across a circuit in bits per second. However, throughput is supposed to measure the transmission of useful application data. When throughput is calculated, overhead bits should be subtracted from the total number of bits that have been transmitted.

## Acronyms

As a long-time acronym hater, I have tried to use them as little as possible. When a term that has an acronym only appears once or twice, the acronym will be noted but never utilized. However, some acronyms really have become the equivalent of words. For example, when referring to digital dial-up service, people do not say "Integrated Services Digital Network," they say "ISDN." Where deemed convenient, the text will remind you of the meaning of acronyms that are common in networking speech but may be unfamiliar to the reader. In any case, there is a list of abbreviations and acronyms at the end of this book that can be consulted when needed.

The LAN Manager's
Internet Connectivity Guide

# 1

# Understanding the Internet

In this chapter we are going to take a brief look at the history of the Internet. We'll examine the role of Internet Service Providers (ISPs) and find out how they link together to form the physical fabric of the Internet. We'll also outline the way in which new applications are introduced onto the Internet and take a look at the typical client/server structure that underlies Internet applications.

# A Brief History of the Internet

The Soviet launch of the Sputnik satellite in 1957 was a wake-up call to U.S. science and technology. In response, the U.S. Department of Defense established the Defense Advanced Research Projects Agency (DARPA) to promote new technologies that could benefit the U.S. military.

High on the wish list was the construction of a communications infrastructure that could survive in a wartime environment. In 1969, a small research network called the ARPANET was built to prove that a new communications technology, packet switching, actually worked. The original ARPANET connected computers at the University of California in Los Angeles, the Stanford Research Institute, the University of California at Santa Barbara, and the University of Utah. Over the next few years, the ARPANET grew into a backbone that enabled several dozen universities and research labs to communicate with one another.

Something very important happened in 1974. Vinton G. Cerf and Robert E. Kahn published a landmark paper that proposed a new set of communications protocols. This paper provided the basis for the development of today's Internet protocols, TCP/IP. The research Internet was born in 1983, when over 100 nodes were switched over to TCP/IP.

In 1986, the National Science Foundation (NSF) took over the research Internet and sponsored a 56-kilobit-per-second backbone. The NSF also sponsored a number of regional networks that provided intermediate connectivity between end user sites and the NSF backbone. This led to the structure that is shown in Figure 1.1.

**Figure 1.1**
The old research Internet.

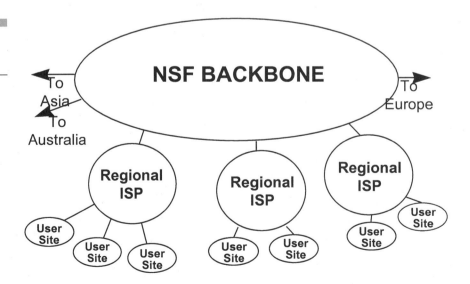

Easy access via regional networks unleashed a period of very rapid growth in the number of institutions that were connected to the Internet. The backbone was upgraded to T1 speed (1.544 Mbps) in 1988 and to T3 speed (44.736 Mbps) in 1991. Currently, some providers are launching backbones that run at hundreds of megabits per second, and a new research network running at gigabit per second speed is under construction.

By the 1990s, some of the regional networks, now called *Internet Service Providers* (ISPs), had started to sell commercial network connectivity. Some ISPs had expanded into coast-to-coast networks.

Nominally, the Internet backbone was dedicated to research and development. But by 1994, many commercial organizations had connected to the Internet, and it was clear that a fair amount of the 10 trillion bytes per month that were traversing the backbone had nothing to do with research and development. A plan to migrate to an unrestricted, unsubsidized Internet was outlined and implemented in short order. The National Science Foundation backbone service was terminated in April of 1995. At this point, the commercial Internet was born.

## What Is Today's Internet?

Figure 1.2 presents a schematic overview of today's Internet. The real picture is hard to draw because the Internet is an interconnected tangle of thousands

**Figure 1.2**
Overview of today's Internet.

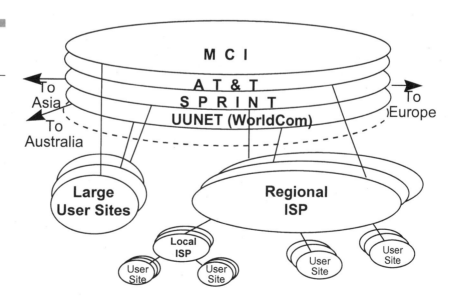

of computer networks. However, it is not difficult to understand if you think of it as similar to the world's telephone network.[1]

Today, several telephone service providers vie for your local and long distance business. The same is true of Internet Service Providers, who wish to sell Internet connectivity to businesses and residential customers. You are able to place telephone calls to locations all over the world because the world's telephone providers are connected together and cooperate in carrying your calls. Similarly, you can set up a connection between any pair of Internet hosts because Internet Service Providers are interconnected and cooperate in carrying your data calls.

The physical Internet has started to look more like the telephone network lately because many telephone companies wish to provide one-stop shopping for all of their customers' communications needs and have added ISP services to their offerings. Familiar names in the telecommunications industry such as MCI, AT&T, and Sprint operate major Internet networks. WorldCom, a global business telecommunications company, entered the Internet arena in 1996 and acquired UUNET, which has been an ISP since 1987.

# Who Are the Internet Service Providers?

Some countries exercise tight control over their ISPs, but in the United States anyone can become an Internet Service Provider. At the time of writing, there were over 5,000 Internet Service Providers.

The telephone companies listed in the previous section are not the only large providers. There are many others, such as BBN Planet, ANS, IBM Advantis, CompuServe, and Netcom. There are dozens of regional ISPs, some spanning several states, and there are hundreds of small ISPs that offer service in a single town or city. In the United States and other unregulated countries, anyone who is willing to invest a few thousand dollars in computers and communications equipment is welcome to plug into the Internet by buying a leased-line connection to a larger ISP and start enrolling customers. In addition, large online services such as America Online and Prodigy offer Internet access to their millions of users.

---

[1]The telephone network is getting pretty complicated too!

# Connecting Service Providers Together

The Internet holds together because large Service Providers have agreed to exchange traffic with one another. The National Science Foundation smoothed the way to the commercial Internet by helping providers to set up some major traffic switching centers called Network Access Points (NAPs). Participants pay a monthly fee to connect to a NAP at speeds ranging from 1.5 to 155 megabits per second. Hundreds of megabits of data traverse a NAP switch each second.[2]

But the NAPs are far from being the whole story. Dozens of other exchange points—where traffic flows from one major provider to another—have been set up in the United States and in other countries. Figure 1.3 shows Internet Service Providers attached to a switching point.

A new, small ISP plugs into the Internet by buying one or more leased-line connections from a larger ISP. Traffic flows to and from the Internet across these leased lines. If the ISP prospers, it can buy a connection to a NAP or other major switching point.

---

[2]The four NSF-sponsored NAPs included the Ameritech Advanced Data Services (AADS) NAP in Chicago, Illinois, the Sprint NAP in New Jersey, the Pacific Bell NAP in San Francisco, and MAE-East, a NAP operated by MFS Datanet (which is owned by WorldCom) in Washington, D.C. Another NAP, MAE-West, was set up by MFS Datanet in the San Francisco area.

**Figure 1.3**
ISPs exchanging traffic at a switching point.

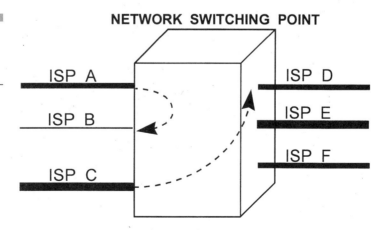

NETWORK SWITCHING POINT

ISP A

ISP B

ISP C

ISP D

ISP E

ISP F

# Computer Names

One of the reasons that everyone from elementary school students to bankers and brokers feels comfortable on the Internet is that Internet computers have names that are easy to guess or remember. Internet systems are assigned names such as:

*www.whitehouse.gov*

*www.ibm.com*

*ftp.apple.com*

*mail1.microsoft.com*

The first two computers are World Wide Web servers, the next is a file transfer server, and the last handles mail that is directed to Microsoft. Computers do not have to be given names that describe a service that they provide. For example, a World Wide Web server at *sunsite.unc.edu* hosts an eclectic array of information ranging from an archive of poetry readings and a historical exhibition called *1492: An Ongoing Voyage* to detailed technical information about several computer operating systems. Fortunately, most organizations assign a name like *www.my-company.com* to a master Web server. The master server can provide pointers to other computers at the site.

# Domain Names

Some technical jargon has been invented to describe the structure of names. First of all, an organization (such as Yale University) chooses a *Domain Name* (*yale.edu*) that conforms to an Internet naming format. Once Yale has registered this name, it "owns a domain." Yale can name the computers in its own domain anything it likes, as long as all of the names end in *yale.edu*. Yale can then divide its domain into *subdomains*, such as:

*english.yale.edu*

*geology.yale.edu*

*physics.yale.edu*

This is convenient because the job of assigning names to computers can be delegated down into the various departments. A departmental administrator has to make sure that each computer in its own subdomain has a unique name, such as:

*keats.english.yale.edu*

*newton.physics.yale.edu*

A complete computer name is called a *Fully Qualified Domain Name* (FQDN).

# Internet Addresses

Although users like to identify computers by name, data is routed to Internet destinations by means of numeric addresses. These addresses are analogous to the telephone numbers that you use to make telephone calls.

Internet (IP) addresses are written as four numbers separated by dots. For example, addresses of some of the computers listed earlier include:

| | |
|---|---|
| *www.whitehouse.gov* | 198.137.240.91 |
| *www.ibm.com* | 204.146.46.133 |
| *ftp.apple.com* | 17.254.0.22 |
| *mail1.microsoft.com* | 131.107.3.41 |

Each of the four numbers in an IP address is in the range 0 to 255. We'll explain this size limit in Chapter 7.

# Internet Automated Directory System

To place a telephone call you enter the destination's telephone number. If you don't already know the number, you look it up in a directory. United States telephone directories are spread out all over the country. If a telephone number is not local, you need to call a special number that connects you to directory assistance at the remote location. For example, to get directory assistance for Phoenix, Arizona, you call 1-602-555-1212.

The Internet directory system translates computer names to computer addresses. Happily, the lookups are automatic. Below, we ask for a login connection to an Internet host named *ds.internic.net*. Note that the second line shows that the name of the computer has been translated to an address.

```
> telnet ds.internic.net
Trying 192.20.239.132 ...
Connected to ds.internic.net.
Escape character is '^]'.
```

```
InterNIC Directory and Database Services

Welcome to InterNIC Directory and Database Services provided by
AT&T. These services are partially supported through a cooperative
agreement with the National Science Foundation.

Users may login as guest with no password to receive help.
. . .
login: guest
```

If you are using a World Wide Web browser such as Netscape Navigator or Microsoft Internet Explorer to connect to a World Wide Web site, you will see messages like the following flash across the bottom of the screen:

```
Looking up www.internic.net
Connecting to 198.49.45.10
```

These messages report your browser's progress as it requests a directory lookup, contacts the remote site, and sets up a connection to the site.

The Internet directory service is called the *Domain Name System* (DNS). The Domain Name System is made up of thousands of cooperating directory servers (called *Domain Name Servers*) spread out all over the world. To get information, your computer will pass a query to a local name server. If it needs to, the local server will automatically communicate with remote name servers in order to get answers for you. Some day the telephone system will become this convenient! Later in this book we will have quite a lot to say about the Domain Name System.

# Routers

The Internet grew from a few thousand hosts to tens of millions of hosts in a decade. This was possible because the Internet is based on Cerf and Kahn's TCP/IP technology. TCP/IP networks are glued together by *routers,* which are special-purpose computers that forward data toward its destination.

Routers provide the total flexibility that supports the Internet's explosive growth. To connect a local area network (LAN) to the Internet, you need to obtain a router, plug it into your LAN, and then connect the router to a circuit[3] that leads to a router operated by an Internet Service Provider (as shown in Figure 1.4). *After you have connected to the Internet, you are part of the*

---

[3]This circuit might be a dial-up line, a leased telephone line, or one of the many other choices that will be described in Chapter 5.

**Figure 1.4**
Connecting a site
router to an ISP
router.

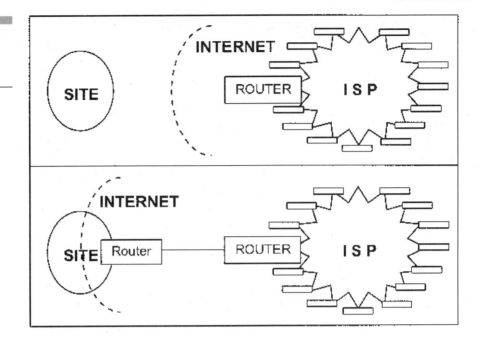

*Internet.* Traffic will flow between your site and the rest of the Internet via your router.

Note that only a fraction of the site in the figure has been added to the Internet. Most sites prudently keep part of their network private. The art of Internet security consists of making sure that the information you share with the Internet actually consists of the information that you want to share (if any).

The link between your router and your Internet Service Provider's routers can be set up using any convenient technology, ranging from a simple dial-up using an inexpensive modem to a high-priced, high-bandwidth leased line. We will describe typical options in Chapter 5.

# Datagrams

Data is sent across the Internet in small chunks called *datagrams*. A header that identifies the source and destination addresses is tacked onto the front of each datagram. A router examines the destination address in order to forward the datagram toward its destination.

# Routing Traffic Across the Internet

It is easy to view Internet routers in action. Below, we show paths that have been traced between a Unix host named *katie.vnet.net* (owned by the VNET Service Provider in Charlotte, North Carolina) and other sites around the United States.

We will use a program called *traceroute*. This program sends three datagrams to each router along the path to a destination. The word "packet" appears in the display below. Packet is a more general term than datagram. In this context, "packet" means "datagram."

The times between sending a datagram message and getting a response back are recorded on the right. Each of the routers has a name that provides a clue as to where it is and who owns it. Each router's IP address appears in parentheses following its name.

In step 2 below, we see that VNET is connected to the MAE-East NAP. Recall that this is a major switching point in the Washington, D.C. area. Many Internet Service Providers connect to one another at this point. Note that the route passes through the MAE-East NAP, across the MCI wide area network, and onto the Yale campus.

```
> traceroute elsinore.cis.yale.edu
traceroute to elsinore.cis.yale.edu (130.132.143.21), 30 hops max,
40 byte packets
 1 rtr-char1.vnet.net (166.82.1.1) 4 ms 3 ms 3 ms
 2 mae-east1-T3.vnet.net (206.80.168.230) 12 ms 12 ms 12 ms
 3 mae-east-plusplus.washington.mci.net (192.41.177.181) 15 ms
   14 ms 15 ms
 4 core2-hssi2-0.Washington.mci.net (204.70.1.213) 16 ms 17 ms 17 ms
 5 border7-fddi-0.WestOrange.mci.net(204.70.64.51) 25 ms 26 ms 23 ms
 6 yale-university.WestOrange.mci.net(204.70.71.110)50 ms 50 ms 49 ms
 7 sloth.net.yale.edu (130.132.1.17) 51 ms 50 ms 54 ms
 8 elsinore.cis.yale.edu (130.132.143.21) 51 ms 52 ms 55 ms
```

The next route crosses the SprintLink backbone to an Internet Service Provider called "TheNet," and then reaches the University of Texas campus network.

```
> traceroute www.utexas.edu
traceroute to www.utexas.edu (128.83.40.2), 30 hops max, 40 byte
packets
 1 rtr-char1.vnet.net (166.82.1.1) 3 ms 3 ms 3 ms
 2 mae-east1-T3.vnet.net (206.80.168.230) 12 ms 12 ms 12 ms
 3 sl-mae-e-f0/0.sprintlink.net(192.41.177.241)109 ms 140 ms 104 ms
 4 sl-dc-8-H1/0-T3.sprintlink.net(144.228.10.41)183 ms 58 ms 41 ms
 5 sl-fw-5-H4/0-T3.sprintlink.net(144.228.10.18) 42 ms 53 ms 125 ms
 6 sl-fw-15-F0/0.sprintlink.net (144.228.30.15) 43 ms 42 ms 55 ms
```

```
 7 sl-uoftx-1-H0/0-T3.sprintlink.net(144.228.135.34) 54 ms 140 ms
   50 ms
 8 ut5-h2-0.the.net (129.117.16.241) 49 ms 46 ms 77 ms
 9 ser2-f8-0.gw.utexas.edu (129.117.20.11) 46 ms 46 ms 64 ms
10 com.gw.utexas.edu (128.83.8.40) 47 ms 46 ms 47 ms
11 homer.cc.utexas.edu (128.83.40.2)46 ms 68 ms 47 ms
```

And finally, we reach a Microsoft Web server by crossing AlterNet (which is the same as UUNET) and then traversing the Microsoft Network. Note that the router at step 6 is slow in preparing its responses to *traceroute*, but there is no delay when it actually forwards messages onward to other routers.

```
> traceroute www.microsoft.com
traceroute to www.microsoft.com (207.68.137.53), 30 hops max,
40 byte packets
 1 rtr-char1.vnet.net (166.82.1.1) 4 ms 3 ms 3 ms
 2 mae-east1-T3.vnet.net (206.80.168.230) 12 ms 11 ms 12 ms
 3 CORE-TR04-AP201.SNFC.grid.net (206.80.180.5) 76 ms 76 ms 76 ms
 4 san-jose3.ca.alter.net (198.32.136.42) 91 ms 91 ms 91 ms
 5 Hssi1-0.AR2.SFO2.ALTER.NET (137.39.100.18) 91 ms 92 ms 93 ms
 6 Fddi4-0.AR1.SFO2.ALTER.NET(137.39.41.129) 274 ms 306 ms 278 ms
 7 Dist1-SF.MOSWEST.MSN.NET(137.39.100.230)108 ms 108 ms 107 ms
 8 msft1-f0.moswest.msn.net(207.68.145.46)109 ms 144 ms 140 ms
 9 www.microsoft.com (207.68.137.62) 112 ms 152 ms 144 ms
```

The *traceroute* program was originally developed for Unix computers, but a version called *tracert* has been bundled with Microsoft Windows 95 and NT and is available from their command prompt. Here is a trace of a path from a desktop computer dialed into Yale's network to the Federal Express World Wide Web site. This path crosses the MCI network to a NAP in New Jersey and then crosses the ANS provider network. Note that the time[4] to the first router and back is quite long. This is because response across a dial-up line is slow.

```
C:\WINDOWS>tracert www.fedex.com

Tracing route to www0.fedex.com [199.81.92.10]
over a maximum of 30 hops:

 1 158 ms 154 ms 158 ms yale-remote-01.net.yale.edu [130.132.57.20]
 2 155 ms 157 ms 156 ms pride.net.yale.edu [130.132.21.1]
 3 158 ms 157 ms 168 ms bifrost.net.yale.edu [130.132.1.100]
 4 183 ms 186 ms 210 ms border7-serial4-4.WestOrange.mci.net
   [204.70.71.109]
 5 187 ms 184 ms 302 ms core2-fddi-0.WestOrange.mci.net
   [204.70.64.49]
 6 190 ms 186 ms 190 ms sprint-nap.WestOrange.mci.net
   [204.70.1.210]
```

---

[4]The Windows *tracert* lists times first instead of last.

```
 7 190 ms 189 ms 222 ms sprint-nap.WestOrange.mci.net
   [204.70.1.210]
 8 377 ms 188 ms 188 ms ans.sprintnap.net [192.157.69.13]
 9 198 ms 200 ms 208 ms h10-1.t32-0.New-York.t3.ans.net
   [140.223.57.30]
10 197 ms 198 ms 203 ms f31.t56-1.Washington-DC.t3.ans.net
   [140.222.56.121]
11 210 ms 206 ms 206 ms h14.t72-0.Greensboro.t3.ans.net
   [140.223.57.17]
12 220 ms 216 ms 223 ms h13.t104-0.Atlanta.t3.ans.net
   [140.223.73.10]
13 225 ms 216 ms 222 ms f0-0.cnss108.Atlanta.t3.ans.net
   [140.222.104.196]
14 320 ms 234 ms 234 ms enss440.t3.ans.net [192.103.73.94]
15 236 ms 238 ms 248 ms 198.83.161.100
16 295 ms 385 ms 292 ms www0.fedex.com [199.81.92.10]

Trace complete.
```

# Internet Applications

Those of us who use the Internet today benefit from the many applications that have been designed, tested, and implemented by Internet researchers. By tradition, new applications are standardized and made available to the public at no cost. Vendors can include these applications in their product offerings without paying royalties.

The result is that when you buy a new desktop computer today, it will include software that enables you to plug into a TCP/IP network and run a diverse selection of Internet applications. You can get free applications by downloading them from the Internet, or you can buy packages of applications for a nominal fee.

The most popular applications are the ones that enable you to:

- Access documents or database information at World Wide Web servers.
- Exchange electronic mail.
- Copy files between computers.
- Obtain a customized daily electronic newspaper or subscribe to news bulletin board groups.
- Participate in "chat" discussion groups or whiteboard conferences.
- Login to a computer across a network.

All of these applications were developed by Internet enthusiasts for use on the Internet. But these applications have proved to be very useful, and have been installed on thousands of completely private networks (*intranets*) all over the world.

If you have a Windows 95, Windows NT, or Unix computer, then the TCP/IP software needed to run Internet applications was bundled with your computer. If you have Windows for Workgroups 3.11, you can get free TCP/IP software from Microsoft. If you have Windows 3.1, there are many vendors that offer reasonably priced packages that include TCP/IP and a big bundle of Internet applications.

Once TCP/IP is installed on your system, you can download new applications from the Internet or buy them in shrink-wrapped form and run them right away. The fact that millions of desktops are ready and waiting has encouraged software developers to follow their imaginations and create applications for whiteboard and voice conferences, live audio and video, three-dimensional explorations, and more.

Extra value has been added to the World Wide Web browser environment by the invention of *Java* and *ActiveX* program language elements. Java or ActiveX application programs ("applets") can be downloaded to desktops when they are needed in order to execute a particular task.

# Role of the World Wide Web

In this chapter, we have devoted most of our discussion to the way that the Internet is physically put together and to the many organizations that provide access to the Internet. Physically, the Internet is made up of computers that communicate with one another, routers that forward their traffic, and lots of local area networks and telephone lines.

Some people think that the World Wide Web is the same as the Internet, but the World Wide Web is just one of many Internet applications. The Web originally was created to enable users to access documents located at many different servers with great ease. Because of the way that the World Wide Web service was designed, it appears to be endlessly expandable. Almost every month someone comes up with a new and exciting extension that makes the Web more useful and more exciting.

# Socket and Winsock Programming Interfaces

Good planning and a lot of effort went into creating the environment that makes it so easy for you to add new communicating programs to your system. The first step was the development of the standard Berkeley Socket Programming Application Programming Interface (API) for Unix. Any program that was developed using the programming calls in the Socket Programming API could run on a conforming TCP/IP system.[5]

A similar programming standard called the Windows Socket Application Programming Interface (or *Winsock* for short) accounts for the ease with which applications from many different sources can be mixed and matched on Windows desktops and can run on top of TCP/IP networking software that has been written by many different vendors.

The actual implementation of the programming interface along with the underlying TCP/IP software is stored in your Windows computer as a bundle of programs called a Winsock Dynamic Link Library (DLL). Once a Winsock DLL is installed, you can add and run any Winsock compatible TCP/IP application programs that you wish, like adding icing to a cake. Figure 1.5 illustrates a few sample applications layered on top of a Winsock interface and TCP/IP software.

Lots of vendors have written Winsock libraries. Each does the same job, but in a slightly different way. To use TCP/IP, your system needs one Winsock. If by accident you install a second one, you will have a mess. It can be just about impossible to sort out the two Winsocks and uninstall one of them. Occasionally application software is shipped with a Winsock that has been included "for convenience." When you install TCP/IP applications on

---

[5]There actually are programming differences across the many different Unix vendor platforms, but application code usually ports fairly easily unless the application is very complex and needs to be optimized for extra-high performance.

**Figure 1.5**

Applications running on top of a Winsock interface.

| Electronic Mail | World Wide Web | Electronic Newspaper | File Copy | Stock Quote Service |
|---|---|---|---|---|
| Winsock Application Programming Interface ||||| 
| Transmission Control Protocol (TCP) |||||
| Internet Protocol (IP) |||||
| Network Adapter Card |||||

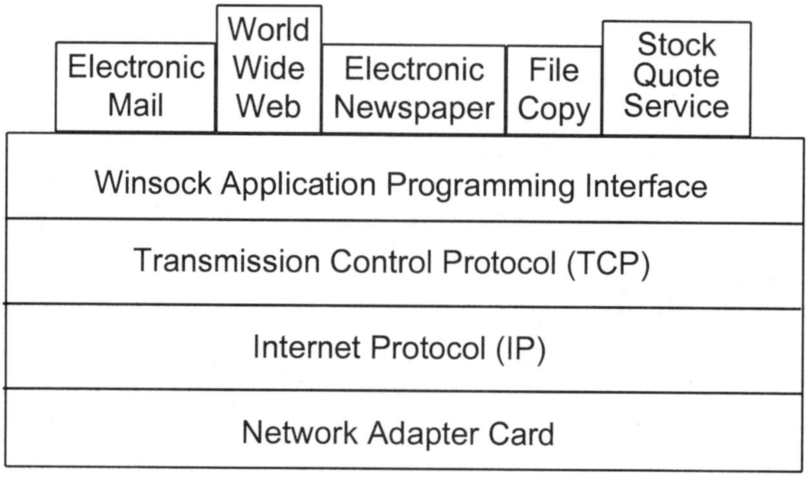

your desktop, make sure that you are not adding an unwanted Winsock. If you have a Windows 95 or NT system and you want to use a third party Winsock instead of the bundled Microsoft Winsock, be sure to remove the Microsoft TCP/IP protocol[6] before installing the new package.

# Clients And Servers

One thing that all Internet applications have in common is that they are based on a client/server model. Familiar examples of clients and servers include:

- A desktop mail application that retrieves electronic mail from a mail server.
- A World Wide Web browser application (e.g., Netscape Navigator, Netscape Communicator, or Microsoft Internet Explorer) that displays information retrieved from World Wide Web servers.

It is important to keep in mind that a user is not the same as a client. A client is a piece of software. When you access a World Wide Web server, your desktop browser client connects to a Web site, retrieves a page of information, and displays it to you. Similarly, you create outgoing mail and receive incoming mail items using an electronic mail client application. In a typical client/server interaction:

- A client application sends a request to a server.
- The server processes the request and sends back a response.

In Figure 1.6, a program in the small desktop system is acting as the client and a program in the large computer is acting as the server. However, it is important to be aware that in the world of TCP/IP, any host can act as a client, a server, or both. It is just a matter of installing the right application software.

It has been a long-standing tradition for the Internet research community to distribute and maintain free TCP/IP client and server software for Unix systems. This was an important factor in the dominant choice of Unix platforms as Internet servers. NT platforms gained acceptance as Internet servers after the release of NT Server version 4.0, which included a generous assortment of free server software.

---

[6]Select **Network** on the **Control Panel**. Then click on TCP/IP bindings and click **Remove**.

**Figure 1.6**
A client request and
a server response.

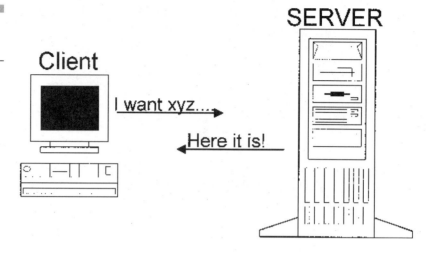

Today, Microsoft Windows 95 products provide some free TCP/IP clients, including:

- The Internet Explorer World Wide Web client.
- A *telnet* client for terminal login emulations.
- A file transfer client, used to copy files between systems.
- A client that sends print jobs to a remote print server.
- A client that sends queries to a domain name server directory.
- A *ping* client, used to check whether a remote system is running.

The Internet Explorer client is featured prominently on the desktop. The other clients are hidden away in the *C:\WINDOWS* directory.

Windows NT Server includes many clients. If you check the *C:\WINNT\system32* directory, you will find clients for telnet, file transfer, *ping*, print job submission, and Domain Name Server queries. Many of the Windows clients are modeled on text-based Unix counterparts.

Windows NT Server also includes a generous selection of server applications, including (among others):

- World Wide Web
- File Transfer
- Domain Name Service
- Print Service

There are, of course, many third party application software packages that you can buy for Windows systems, and these contain a diverse array of clients and servers.

# Evolution of Internet Technology

The Internet was developed by a method that is unique in the history of human endeavor. The hard work of introducing and perfecting protocols was, and still is, done by volunteers who are organized as the *Internet Engineering Task Force* (IETF). The people who created and operated the Internet also were fervent end users. Applications were designed by people who wanted to have and use those applications. The applications were improved by other designer-users. The result has been that new ideas, tools, and services have arisen, evolved, and spread at an unprecedented rate.

## REFERENCE INTERNET DOCUMENTS

A feature that has contributed to the rapid spread of Internet technology is the fact that all of its documents are online and free. The most important documents, called *Requests For Comments* (RFCs) currently can be retrieved from a directory called */rfc* at *ds.internic.net.*[7] This site, called InterNIC Directory and Database Services, is operated by AT&T and was sponsored by a grant from the National Science Foundation.

If you wish to retrieve documents from this site, it is a good idea to start with the index file, */rfc/rfc-index.txt*, which contains titles of all of the documents.

---

[7] RFC documents can be retrieved via file transfer (*ftp://ftp.internic.net/rfc/rfcxxxx.txt*). Documents also can be accessed via the World Wide Web (*http://www.internic.net/*).

# 2

# Doing Business on the Internet

The introduction of the World Wide Web application to the Internet sparked the great Internet gold rush of 1995 and 1996.

Many organizations have raced to plant a Web site on the Internet and most have benefited from taking this action. A Web site is a relatively inexpensive way to provide information 24 hours per day, every day of the year, to an audience that spans the globe. However, many sites do not take advantage of the potential that a Web presence offers to help to market products, improve customer service, reduce costs, and expand into new business areas.

In this chapter, we will examine several levels of Web service and look at examples of companies that are using their Web sites to achieve business goals. We'll provide some tips that make a Web site more visible, help prospective clients to find your organization, and enable users to reach the information that they need quickly. Finally, we will describe several all-too-common bad Web site practices that frustrate customers, choke the communications lines leading to your Web site, and cripple server performance. Chapter 8 and Chapter 13 will present more detailed technical information relating to Web server performance.

Before we start, let's survey some of the terminology that has grown up around the World Wide Web. Chapter 13 is devoted to Web issues, but we cannot wait until then to review the basics.

# Web Terminology

Anyone who has used one of the popular desktop applications such as Netscape Navigator or Microsoft's Internet Explorer knows how easy it is to connect to a World Wide Web server and retrieve information. Most of the time, all that you have to do is point and click.

## Browsers, Pages, and Links

Recall that a client application (such as Navigator or Explorer) is called a browser. The documents that you retrieve are called *pages*. Web pages are special because they contain phrases that you can click in order to *link* to a new page.

Document links are nothing new to you if you have ever used Help at a desktop computer. You know that if you choose one of the underlined items in Figure 2.1, you will see a new page of help information that will discuss the

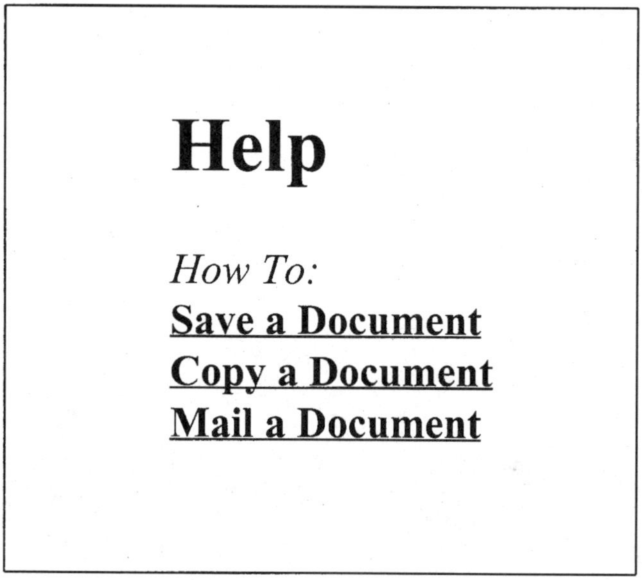

**Figure 2.1**
A Help page that contains links to other pages.

# Help

*How To:*
**<u>Save a Document</u>**
**<u>Copy a Document</u>**
**<u>Mail a Document</u>**

topic that was chosen. Each help page actually is a file that will be retrieved from your hard disk or from a CD-ROM. Help pages sometimes contain pictures, sound, or even film clips.

## Hypertext Pages

Pages that contain links that can take you to other pages are called *hypertext* pages. A file that contains a hypertext page includes extra hidden information for each underlined or specially marked phrase. This hidden link information points to the file that should be loaded when a user clicks on the phrase. Hypertext pages also contain hidden instructions that describe how text should be formatted, and pointers to pictures or sounds that should be presented with the page.

## Hypertext Tags

Web pages are hypertext pages. Each page is a simple text file. Hidden information is enclosed in angle brackets and is labeled with a tag letter or phrase. For example, the <B> tag in the example below indicates that text that follows should be presented in boldface type, while the </B> tag marks the end of boldface text:

```
<B>This really is important!</B>
```

## Hypertext Markup Language

The set of tags that are used for Web pages form the *Hypertext Markup Language* (HTML). HTML tags define section headers, describe tables, indicate where pictures should be placed, and describe links to other pages.

## Uniform Resource Locator

A *Uniform Resource Locator* (URL) is the key information that stands behind a Web link. Here are some examples of URLs:

*http://www.yahoo.com/*
*http://www.bbn.com/index.html*
*http://cirrus.sprl.umich.edu/wxnet/*

- The first URL says, "Get me the default page at host *www.yahoo.com.*" A site's default page also is called its *home page*.
- The second URL says, "Get me the page called *index.html* at host *www.bbn.com.*"
- The third URL says, "Get me the default page in directory *wxnet* at host *cirrus.sprl.umich.edu.*"

Now that we have a basic vocabulary in place, we are ready to move on and look at some typical Web server roles.

# Marketing Brochure

The simplest type of Web site simply presents a company's marketing brochure. This usually includes information about a company's location(s) and size, its annual report, and brief descriptions of its major product and service offerings. A site of this type does not make full use of the potential of the Internet, but it does deliver basic information about a company to an audience that not only is very large, but also is international.

# Product and Company Information

Computer and networking technology companies were the groundbreakers for sites that put all of their latest product information online. In today's technology environment, product life cycles are short and change is constant. Printed matter cannot keep up with the dizzying pace of innovation and change. Even an organization's own sales representatives have a difficult time keeping up with their product offerings.

A Web site is an ideal place to publish product announcements, news releases, product information, technical specifications, and answers to commonly asked questions. For many companies, their Web site has become a primary information source for their own staff as well as for their customers.

# Customer Support

Just about any service that depends on customers calling an 800 number and requesting information is a candidate for installation at a Web site. Customers usually are quite capable of entering queries into an information database themselves instead of having a staff member do it for them. Every time that a visit to a Web site replaces an 800 number call to support personnel, an organization saves on costly telephone bills (those half-hour wait times can really add up) and conserves valuable staff resources.

Online support also presents a company with a marketing opportunity. While customers are visiting a Web site to get hard information, they can be notified of new products and services.

# Sales

Software, computer equipment, and catalog sales companies are right at home on the Web. They have been joined by food and wine stores, gardening suppliers, clothing catalogs, and more. For some, sales are booming. The rollout of *Secure Electronic Transaction* (SET) applications during 1997 and 1998 is expected to increase the volume of online sales significantly. SET supports safe credit and bank card purchases on the Internet.

# Examples

In the sections that follow, we'll point to some positive role models and describe actions that you can take to make your site easy to find.

# Aetna

If you visit the Aetna site (*www.aetna.com* ), you'll find that Aetna presents far more than product and marketing information. Aetna's site has helped the company to cut down substantially on the time that their staff members need to spend on the telephone taking care of customer support needs.

Many people worry about investing for their retirement, but are not willing to talk to a sales representative—either because they do not wish to take the time or because they do not wish to be subjected to sales pressure. At Aetna's site, visitors can analyze their retirement requirements and capabilities using simple forms, without having to talk to a salesperson.

Customers who already have a retirement account with Aetna have secure online access to up-to-date information about the current value of their accounts. This improves customer service while cutting down on costly phone support.

The health information section enables members of Aetna health care plans to find participating doctors, hospitals, and other medical services. This section also provides tips that help to keep subscribers healthy by publishing lots of free online health information pamphlets.

The Aetna Web server promotes the company's enterprises in every way and is solidly business-like. All information can be viewed when accessing the site in text-only mode. Every page contains useful data. The site is well organized and easy to navigate.

# American Airlines

The American Airlines site (*http://www.americanair.com/*) provides another example of making good use of a Web presence. Their Web page is used for air travel transactions.

Customers can check departure and arrival times. Every customer who uses this service saves American the fee for an 800 number call. Customers can use a secure form to purchase or reserve their own tickets, which adds to American's profits by cutting out a travel agent's fee. American also fills empty seats on upcoming flights by offering last-minute fare specials that can be ordered via an interactive form.

# Federal Express

Federal Express (*http://www.fedex.com*) surprised their competitors by being the first delivery service to enable customers to track their packages online at a Web site. They offer another inducement to ship Federal: free software that prints usable labels, complete with bar codes, and keeps a record of the packages a customer has sent.

The Federal Express site enables customers to register for an account number, schedule pickups, or find a drop-off location. If something goes wrong with some shipment, customers can use an online form to report delivery problems and arrange for credit on the transaction.

## Security First Network Bank

The Security First Network Bank (*www.sfnb.com*) is a remarkable Internet enterprise. This Federal Deposit Insurance Corporation (FDIC) insured bank operates on the Internet, using secure transmissions to safeguard its customers' financial transactions. The bank provides the usual bank services, such as checking and savings accounts, ATM and bank cards, electronic payments, money market investments, CDs, and stock market news. Access is via an ordinary World Wide Web browser.

The "Bank" is open on the Internet 24 hours per day, 365 days per year. It provides electronic mail and telephone support to its customers. Since Security First Network Bank does not have to pay for an imposing building, marble lobby, furniture, etc., this bank operates very efficiently. Its low overhead enables the bank to provide low-cost services to its customers.

## Would You Buy a Used Car from the Internet?

Used car revenues add up to $350 billion dollars per year, and an increasing number of used car sales are being brokered across the Internet. Internet used car sites help customers to find the model and features that they are looking for. In some cases, groups of dealers have joined together to provide warranty support so that a car bought from a remote vendor can be serviced locally. For some examples of how this business works, see:

> *http://www.priceautooutlet.com/edmunds/*
>
> *http://carpoint.msn.net/*
>
> *http://www.autoresponse.net/*

## Specialty Stores

You do not have to be a corporate giant to make good use of the Web. Many food and wine specialty stores are among the smaller businesses that have made a success of their Web sites. The legendary Virtual Vintner site (*http://www.virtualvin.com*) attracted customers by answering customer questions about wines ("Ask the Cork Dork") and providing access to fine and sometimes hard to obtain vintages. The Virtual Vintner has grown into a popular wine and gourmet food center. Customers can safely and easily order merchandise by filling in an online form.

As you might guess, businesses that sell goods that have some unique characteristic or are difficult to obtain outside of a particular region can do well on the Web. For example, the Cook's Garden's Web site (*http://www.cooksgarden.com/*) has helped this high-quality vegetable seed and plant catalog business to grow from a small retail stand to a major distributor.

## PointCast

At some sites, information *is* the business. Innovator PointCast has created a business that builds customized electronic newspapers for its clients. Users choose exactly the news that interests them, and there is a broad choice of sources. According to PointCast:

> PointCast broadcasts national and international news, stock information, industry updates, weather from around the globe, sports scores and more from sources like CNN, CNNfn, *Time, People* and *Money* magazines, Reuters, PR Newswire, BusinessWire, Sportsticker and Accuweather! Even your local newspaper will be on PointCast—*LA Times, New York Times, Boston Globe, San Jose Mercury News* and more.

## Recipes for Success

The sites that have been described have several features in common. They provide real benefits, both to their customers and to the businesses that sponsor them. In all but one case,[1] users can access the sites in simple text-only mode and view all information needed to navigate the sites, find and retrieve information, and make purchases. Glitzy images, special technology,[2] and gimmicks do not play a role in their success (although high-tech features

---

[1]PointCast provides special desktop software that operates in the background to periodically download the customized newspaper, headlines, and breaking news.

[2]Such as Java or ActiveX.

might be important to a site that makes its revenues through advertising and wishes to attract visitors who are looking for entertainment).

As you can see, it is a good idea to look at some successful high-quality sites while designing your own Web site. This can stimulate ideas for new ways to do business as well as provide models for presentation. Currently, good sites are easy to find through the *Lycos/Point Top 5% Web Reviews*. At the time of writing, these can be reached at:

*http://www.pointcom.com/categories/*

This site has reviews of well-organized Web sites (called the Top 5% by Pointcom), and provides links to the sites. Some of the Pointcom sites get by on good looks rather than on successful business activity, but many are well worth a visit.

# Free Advertising for Your Pages

The Web is a vast place, but you can take actions that will improve the chances that prospective clients will find their way to your site. The most important task is to do your homework so that your pages turn up as highly-rated answers to appropriate queries at the major Web search sites.

If you happen to have a Netscape browser, you can view a list of some of the important search sites (such as *Yahoo, AltaVista, Excite, Lycos, Infoseek,* and *Four11*) simply by clicking the Net Search directory button. In any case, you can connect to Netscape's search site page at:

*http://home.netscape.com/home/internet-search.html*

You probably already have carried out searches at some of the sites that are listed.

# Building Internet Search
# Site Databases

How does an Internet search site get its information? Strange as it may seem, a robot program visits Web servers all over the Internet and indexes all of the words[3] on each page that it finds. The words are weighted in importance de-

---

[3]Of course, words like "by," "for," "it," etc. are omitted.

pending on their placement (such as appearing in a page title or in headers) and on their frequency. The weighting factors depend on the specific type of indexing software that is used at each search site and on how that software has been configured.

# Registering at a Search Site

Internet search servers allow you to register so that your site will be included in their indices. There usually is a link to a registration form on the search server's home page. Figure 2.2 shows part of Yahoo's registration form.

When registering, you often can add valuable information about your site. By registering:

- You assure that your pages will be found and indexed.

- You provide a short abstract that is displayed as part of the search response.

- You provide keywords and category information that help your site to be included on relevant response lists.

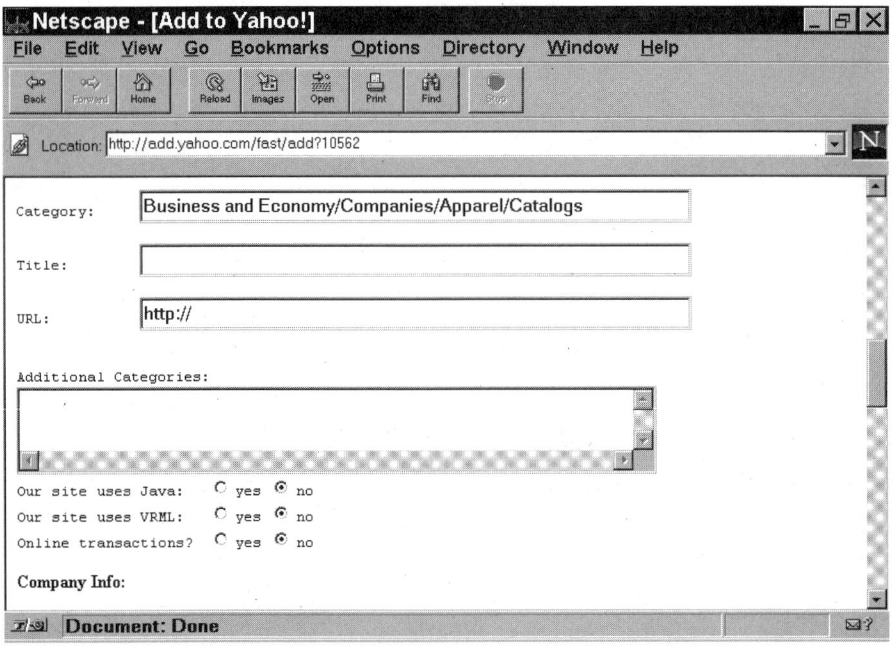

**Figure 2.2**
Registering an apparel catalog company at Yahoo.

```
Netscape - [Add to Yahoo!]
File   Edit   View   Go   Bookmarks   Options   Directory   Window   Help
Back  Forward  Home  Reload  Images  Open  Print  Find  Stop

Location: http://add.yahoo.com/fast/add?10562

Category:        Business and Economy/Companies/Apparel/Catalogs

Title:

URL:             http://

Additional Categories:


Our site uses Java:     ○ yes  ⊙ no
Our site uses VRML:     ○ yes  ⊙ no
Online transactions?    ○ yes  ⊙ no

Company Info:

Document: Done
```

Your abstract statement is very important. For example, the responses to a search for the words "audio equipment" returned a list of companies whose comments included statements such as:

"We design and build music recording studios and post production facilities all over the world."

"We manufacture high quality loudspeakers that offer exceptional sound reproduction."

"Supplier of audio, video, and television accessories, cables, and components."

Statements like these help a user to choose a relevant site. It is a good idea to include search keywords that relate to your site in your abstract statement. This should improve your score when users search on those words.

For best results, you should take the time to visit the search sites that you feel are especially important, read their instructions, and register at each individually. The procedures will differ slightly for each of them. For example, Yahoo's site is organized into subject categories and you are allowed to register in two selected categories. To do this, you need to navigate to the screen for a category and fill out a form there.

# Using a Registration Service

There are dozens of search services on the Internet and you probably do not want to take the time to track them down, visit each one, and go through their registration procedures. Fortunately, when there is a need on the Web, there usually are companies that are ready to satisfy that need. For example, Submit It! (*www.submit-it.com*) provides a registration form and a check-off list for hundreds of sites at which you might wish to be registered. They will perform a few registrations at no cost, but offer a more extensive registration service for a modest fee.

There now are dozens of services that promise to promote your site through registrations, advertisements, planting links, and other methods. See the lists at:

*http://www.yahoo.com/Computers_and_Internet/Internet/
World_Wide_Web/Announcement_Services/*

*http://www.yahoo.com/Business_and_Economy/Companies/
Internet_Services/Web_Presence_Providers/Announcement_Services/*

# Getting to the Top of the List

In a brochure, magazine, or newspaper advertisement, you might depend on recognition of your company's name or of your graphic logo to attract readers. But in order to be included and prominently displayed by a search list, your text must contain the key words and phrases that people might search for.

If you sell water pumps, then your abstract, keywords, and one or more major headers should contain the words "water pumps." Here is the top response to a search for "water pumps."

```
Global Water Pump Expo - virtual expo of suppliers, products, and
services for pumps and related products.
```

Note that the words "Water Pump" were right up front in the page title.

# Providing a Search Tool

Next we are going to see how installing your own search tool can enhance the quality and efficiency of your site. Many of the search tool products on the market are outgrowths of the software that was developed for the Internet search sites.

Smart companies help visitors find the information that they need. This makes patrons happy. It also should make the host company happy because visitors are not tying up the site by retrieving page after page of irrelevant information as they wander toward their objective.

Let's take a look at an example. I am interested in finding out about a new secure Web server called the Virtual Vault that Hewlett-Packard is marketing. Clicking Search HP at the bottom of the Hewlett-Packard home page (*http://www.hp.com/*) causes the search box in Figure 2.3 to be displayed.

The response to a search for "Virtual Vault" starts with:

---

We found 1772 documents that matched the words in your query: Virtual Vault. The best 30 matches are listed.

100     <u>VIRTUAL VAULT TECHNOLOGY</u>

. . .

---

**Figure 2.3**
A search form at the
Hewlett-Packard site.

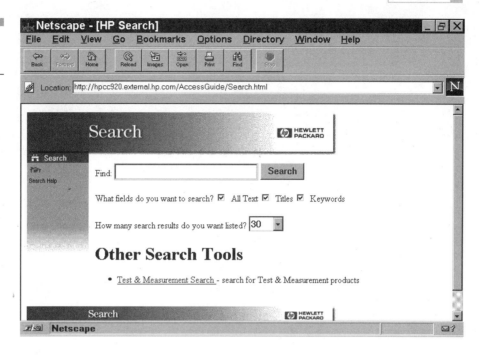

**Figure 2.3**
A search form at the
Hewlett-Packard site.

This paper is just what I was looking for. I now can view this paper by clicking on its title. We'll look into search tool features later in this chapter.

# Search Tool Features

A good search tool is an indispensable part of any Web site. There are many search tools available and choosing one should be done with some care. Desirable search tool properties include the ability to:

- Search for phrases (such as "stock market") as well as individual words.
- Search for words and phrases that appear near one another.
- Use wild cards (such as "chem*" or "chapter?").
- Search by properties such as a document's title, author, or date.
- Use (Boolean) operators such as AND, OR, and NOT.
- Search text, hypertext, or word processing documents, Adobe Systems Inc. Portable Document Format files, or other application data.
- Allow the user to choose how many answers will be returned in each response page.

A search tool returns titles of documents that contain the given words and phrases or satisfy other criteria (such as the name of the author or creation within some time period). The documents are sorted and the ones that match best are at the head of the list.

A bald title often is not all that helpful in deciding what a document really is about. Good search tools provide a short abstract along with each title. An abstract is very helpful to the end-user.

Some search tools provide "concept searching." One way that this works is that a thesaurus lookup expands the user's original list of keywords. For example, a search term such as "dog" would be broadened to include "canine" and "hound." Sometimes this is useful. At other times, it can cause a search to go off track and return a lot of garbage. Therefore, it is helpful to have a way for users to specify that their precise words must be present in the documents.

Some products use a method that does a statistical correlation between documents that match the original search criteria and others that contain substitute words. This makes it far more likely that a concept search will produce relevant documents.

## How Search Tools Work

Search tools create a list of document titles in response to words or phrases that are entered by a user. The user can view any document on the list by clicking on its title.

The key to swift searches is that all of the words in the set of accessible documents are pre-indexed. The searchable documents may be stored across several directories. Some search tools can index documents that are spread across several computers. One way to do this is to preconfigure a list of directories at the various computers and set these up as "virtual" directories that are part of a Web server's resources. The other way mimics what the Internet search tools do, that is, use a robot program that visits a list of servers on a regular basis, indexing and re-indexing their public documents.

## Efficient Searches

A search tool is not helpful if it does not provide what the visitor is looking for. Customers will be driven away by:

- A list of irrelevant selections.
- A list of answers that contains the same document title repeated many times.
- Pointers that lead to error messages because the document has been moved or deleted.

Someone will have to put in some effort up front in order to smooth the way to good search results. Here are some tips worth considering:

- Organize your archive so that users can optionally confine a search to a subset of documents that are most likely to be relevant. For example, you might have separate categories such as corporate profile data, product descriptions, price and delivery information, and customer support information.
- When you create your documents, keep in mind that the title, headers, and first paragraph will be important search elements. Think about keywords that a user might enter when looking for this document, and make sure that these keywords appear in the important positions.
- Add abstracts and keywords to documents and to other items (such as spreadsheets or image files) that you wish to make available.
- Regularly remove obsolete material from your document database.
- Schedule frequent updates of your index so that links lead to current documents.
- Although you may wish most searches to be launched from a very simple form that provides one box in which users enter search words, it is a good idea to create additional forms that allow users to focus their searches—for example, by:
  - Stating where the search term is located (e.g., in the title of a document).
  - Choosing a publication time span.
  - Indicating the type of item that is sought (e.g., price information or a software patch).
- Think about what your customers may need to find, and test your indexer by running many sample searches.

Many of the search tools that are available today include powerful tools for customizing the way that searches are carried out. It takes some time to learn all of the ins and outs, but the payoff for this effort is that customers who get what they are looking for quickly will appreciate and respect your organization and will place fewer telephone calls to customer support. An added bonus is that when you meet customer needs efficiently, your site is freed up to serve additional customers.

# Search Products

This is a lively and competitive product area in which offerings are changing and improving at a brisk rate. We'll discuss a few popular search/index products below. The intention is to provide an idea of some of the capabilities that are available, not to recommend these products exclusively. It is not possible within this limited space to cover all of the good products or even all features of the listed products.

## Microsoft Index Server

At the time of writing, Microsoft's Indexer is a free add-on for NT Server. (You can download it from *http://www,microsoft.com/*.) In addition to indexing simple text documents, this search tool can index text and properties information from Microsoft Office files. The indexer has powerful customization capabilities that can be used to create an effective search environment.

Microsoft's Indexer creates abstracts for Web hypertext documents that include major headers in the document and the first few words of the text. Below, we see the result of a search for the word "security."

Microsoft Index Server Guide: Security
*Abstract:* Security. Catalog Directory Access Control. Access Control for Web Pages. Authentication. Remote Virtual Roots. Administration Over the Web. Configuration Note. Space. Previous. Next.
This section discusses the security administration of Microsoft Index Server. In order to maintain a secure site without ...
   *http://198.207.177.32/srchadm/help/sechelp.htm - size 10,485 bytes - 8/5/96 6:19:06 PM GMT*

## Excite for Web Servers

At the time of writing, the Excite product is free. (You can download it from *http://www.excite.com/*.) Excite can index text and HTML files. The Excite search engine runs on NT and on a range of Unix systems. By the time this book is in print, Excite should be available for Macintosh computers. Excite supports concept searches that are not simply Thesaurus-based. Excite uses statistical correlation techniques to ensure that documents containing alternative words are relevant.

Excite also groups result lists. For example, if a user searched simply for "market," documents relating to the stock market would be grouped separately from documents about market gardening. In spite of this extra functionality, Excite is a lean, fast search tool.

## Livelink Search

Livelink (from Open Text Corporation, *http://www.opentext.net/*) is a powerful search product that can index many types of files including text, HTML, Standard Generalized Markup Language (SGML),[4] Microsoft Office products, Adobe Portable Document Format, and others. Information from multiple Web servers can be indexed using an optional robot component. The product is highly configurable, allowing a publisher to offer options such as:

- Allowing users to choose exactly where in a document they wish to search (Anywhere, Summary, Title, First Heading, or URL).
- Searching for words or phrases joined by AND, OR, BUT NOT, NEAR, or FOLLOWED BY.

Livelink currently runs on NT, Sun Solaris Unix, and HP/UX Unix. (Currently, a Netscape Commerce Server and auditing and billing modules are included with the product.)

## Catalog Server

Catalog Server (from Netscape Communications Corporation) can index many types of files including text, HTML, Microsoft Office, Adobe Portable Document Format, Corel WordPerfect, and Lotus Development Corporation's AmiPro. It can index information from sources spread across multiple computers. A robot information gathering technique is used. The product makes it easy to organize information by subject hierarchies and is customizable.

Catalog Server runs on Windows 95, Windows NT, and several popular Unix platforms. Catalog Server is a commercial product, but you can download a trial version from *http://home.netscape.com/*.

---

[4]The first version of HTML was based on a subset of SGML, which is an international standard for representing formatted text in electronic form.

### Verity's Search '97

The muscular Verity search engine can serve as a middleware connection to database and application data in addition to indexing the usual text, HTML, and office application sources. It supports concept-based searching, Boolean operator terms, proximity criteria, and field criteria (such as "author").

The product is highly configurable, and it can index multiple Web servers and other information sources spread across many platforms. Search '97 is available for NT and several Unix platforms. Information is available at *http://www.verity.com/*.

# Who Designs Web Pages?

Many Web page design firms are outgrowths of marketing companies that have specialized in print media in the past. Unfortunately, many continue to do their work as if they were creating paper brochures that someone can thumb through quickly.

The Internet is a very different medium. Most people want to find specific information and they want to find it fast. When dial-up users access the Internet at modem speeds, they do not have time to leaf through glossy pictures that take minutes to download. Sometimes the process is so slow that the user gets angry and frustrated and just gives up, never to visit that site again. The Web design "experts" have converted a current customer or a sales prospect into someone who is hostile to your company.

It is not just users with slow modems who suffer from the results of poor design. Pages that are packed with graphics and special effects will bog down your Web server, cutting back drastically on the number of clients that it can handle each hour. Remember that while one user with a slow connection is waiting for a download, another user may be denied access because the server is at its peak load. Chapter 8 and Chapter 13 include detailed explanations of Web design practices that can turn a peppy server into a plodding drudge.

Web sites can be designed so that they meet the information needs of their visitors very efficiently. In the next few sections, we'll describe some of the don'ts and do's.

# Bad Web Site Practices

We are going to look at several all-too-common practices that slow Web server performance down to a crawl and drive away customers. But first, let's consider why these things happen. It is possible that you are encouraging your Web developers to build a bad site by rewarding them for doing the wrong things.

Before you approve of complicated, purely decorative graphics, be sure to test them on a dial-up connection. You may discover that you have chewed up your mouse cord in frustration by the time the graphics have arrived.

Many people love to brag about the number of "hits" that their site has per hour or per day. But keep in mind that a hit corresponds to a file that is downloaded, and every icon, button, and image is in a separate file. If you equate hits with success, your Web developers may fatten up your site with extra items that help to swell the tally, or they may attract casual visitors, who have no interest in becoming your customers, by serving up some hot technology tricks.

Hits, or even user counts, are poor measures of success. As we have seen in earlier sections, a Web site can be a potent business tool. You want to attract the right people and you want them to accomplish their business as efficiently as possible—for your benefit as well as theirs.

# A Page that Introduces the Home Page

More and more sites seem to be featuring a "cover" page containing a graphic design and a link to the "real" home page. This may be a carryover from the design of print brochures, which have a pretty cover that has some pictures and the name of the company. A more sinister reason may be a desire to beef up the number of hits by forcing users to access one or more dummy pages before they can get to any information.

Users who have to wade through extra introductory material are not going to be happy. Suppose that you had just waited in a line at a bank and at the end were handed a picture and a pointer to the "real" line. I doubt that you would be a happy customer, and you might wonder why your bank wanted to use its resources to pay people to create useless pictures and hand them out.

# Lack of Organization

Some sites contain pages that meander from topic to topic. You need to plan the architecture of your site just as you would plan the design of a new building. At a successful site, information content usually expands pretty quickly. If you do not have an overall design, you will end up with a labyrinthine mess.

Creating and maintaining a site index will not only help visitors, it also may help you to discover ways in which the material at your site can be reorganized in a more logical fashion. Your home page is a very good place to put your site index. If there is something that prevents you from doing so, then the home page should prominently display a text pointer to the site outline.

As we already have noted, sites that hold a lot of information should provide a search capability. Getting searches to work efficiently takes some manual effort and tuning. Make sure that searches for keywords related to hot topics return the right titles. It is a good idea to gather words and phrases that come up in conversations between your customers and your sales and support people and use these in test searches.

# Too Many Short Pages

Once a user has found a document, the user ordinarily wants to read it. Many Web site authors become HTML slaphappy and break ordinary text documents into lots of short HTML snippets that have to be laboriously retrieved, one small piece at a time.

This increases the hit count, but it causes wear and tear on the Web site, the network, and most of all, the reader. Downloading an entire document makes far more efficient use of resources. A reader who wishes to look up something that appeared in an earlier chunk of a fragmented document has to back up through an odd assortment of pieces, performing a search within each piece.

# Large Image Files

"A picture is worth a thousand words." The average length of the words that I have written for this chapter is 5 characters. So we could translate the statement into, "A picture is worth 5 thousand bytes." Before putting a picture that is 40 or more kilobytes onto a page, think about the user, the server, and the network time that will be expended in transmitting it.

Often, by investing a little work, the color and design of a graphic can be adjusted so that the size of a graphic file is reduced drastically while the appearance remains good and the content is preserved. If a large graphic file really is needed, the recent Portable Network Graphics (PNG) format[5] is more efficient for big graphics than the older Graphics Interchange Format (GIF)[6] format.

## Too Many Images

Putting pictures, icons, and logos onto a paper page can make it attractive, lively, and interesting. A few images may, in fact, enhance the content of a Web site. But loading Web pages with graphic data will slow downloads to a teeth-grinding crawl.

## Requiring Images

Too often, users who choose not to download images are confronted with little or no content. Some sites provide a link from a site's home page to a text-only version of the page. This is better than a blank page, but why should a visitor have to perform two downloads instead of one?

There are many webmasters who manage to create pages that successfully convey all of their important information in text and make sure that all links from that page are visible whether the pages are viewed with or without their images.

Images sometimes are important. A client choosing a bouquet of flowers will want to look at a picture of the bouquet. However, a text description or a small thumbnail image should lead the client to click on that picture. The customer should not have to download twenty big graphics in order to locate a specific one to view.

---

[5]See *http://www.w3.org/pub/WWW/Graphics/*.

[6]GIF is a proprietary format defined by CompuServe Information Services. It uses a compression method patented by Unisys Corporation.

# Using Graphic Links

Some sites get it almost right. The text content provides coverage of the information and all graphics are labeled. However, important links may be hidden behind clickable images, graphic buttons, or icons. The problem is that a separate graphic file has to be downloaded before the user can click on it and select the link. Specifically, the user has to:

- Click on the text in the image area.
- Wait for the image to be retrieved.
- Click again to access the link.

Remember that you are not just wasting the user's time. You are tying up your site with a totally inessential file retrieval session. The solution is very easy: add a text link.

# Requiring Java or ActiveX

Today, many Web sites attempt to download Java or Microsoft's ActiveX programs to users' desktops. Some sites have gone to extremes and, without these programs, offer nothing but blank pages. This is definitely not the route that you should follow today if you want to get some serious work done on the Internet.

What causes the webmaster craze for Java? Java is a programming language that is the answer to a programmer's dream. It is possible to wrap a Java execution environment into Web browsers and servers. The result is that Java programs can be downloaded to browsers on any type of desktop system, or added to Web servers running on any type of computer platform, and will run there. This is a boon to developers who wish to create portable applications.

ActiveX programs run on Windows desktops, and hence add extra application functionality to browsers that run in a Windows environment. Small Java or ActiveX programs are called *applets*.

Java is a real programming language and working with Java requires programming skills and special training. In contrast, a scripting language is similar to a list of commands written in English, and is a lot easier to use. Javascript is a scripting language that complements Java and can be used on desktops and at servers. Javascript commands can be included in "smart" HTML documents.

In any case, if a user's browser accepts Java or ActiveX, then programs can be downloaded to the user's desktop and will be executed there. ActiveX programs have the ability to damage user files. There have been great efforts to make the Java execution environment safe, but at the time of writing, Java still has the potential to cause trouble. Prudent Internet users who are not enthusiastic about letting strangers download executables onto their desktops turn Java and ActiveX off.

The potential security threat posed by software downloaded from a site run by a complete stranger has alarmed many companies that have responded by configuring their firewalls to block all Java and ActiveX from entering their sites. Later, we'll explain how a process called *certification* will be used to create bonds of trust that make it practical to download Internet applets and scripts.

Today it does not make good business sense to prevent customers who are simply exercising normal prudence from viewing your information. It is very important to create pages that display *all* essential information in the text-only view of the page.

# Bad Hypertext Links

If you run an active site, it probably gets reorganized pretty often. Sometimes when files are moved around, the links pointing to the files are not corrected. Too often, users select an item that says something like "Click here to find out about our terrific new technology" and get a "not found" error message.

There are tools that will automatically step through all of your links and make sure that they are valid. For a list of tools, see Thomas Boutell's World Wide Web frequently asked questions document:

*http://www.boutell.com/faq/*

There are many useful frequently asked questions documents on the Web and they provide information about an unbelievably wide range of topics. These documents are called "FAQS" for short.

# Adding Special Effects

Watch out for high-tech special effects. Webmasters like to keep up with the latest technologies and often use them to the point of overkill. Do you want to attract lots of people who visit your site to see a clown that performs somersaults? How many of these visitors will be prospects or customers? And if you are not selling compact disks or show tickets, it is a good idea to omit the background music. You can always provide a button that music lovers can push if they want sound.

## Silly Frames

What is on a browser screen? At the top, there usually is a title bar, a row of menu headers, some toolbar buttons, and a location box. At the bottom, there is a status bar and probably a toolbar that shows what programs are open. Your working area is sandwiched in the middle.

So of course, someone came up with the idea of breaking this tiny area into three or four frame areas, each big enough to show about one teaspoonful of information. Frames clutter a screen, force users to scroll through text that is displayed in tiny windows, and often are just plain inconvenient. One rule that webmasters need to master is, "Just because you can do it does not mean that you should do it."

There is one use of frames that can be helpful. Some Web designers display a master index for the site in a stable frame at the left. The space to the right can be pulled to the left to cover up this index when it is not needed, but it can be uncovered for a quick overview of the site.

## No Pointer to Home

Someone may leapfrog onto one of your pages via a Web search or a pointer from another site. People occasionally reach some interesting information at your Web site by using a search tool that takes them to a file that is not on your site's main Web computer. The user may be wondering, "Where am I?" and "Whose site is this?" and "Is there more like this here?" Putting a pointer to home at the top or bottom of each page allows you to reel in a fish that has landed in your net.

## One-Way Conversations

In any endeavor, nothing is more valuable than getting information from your customers. Fortunately, there is something about the Internet experience that makes many people willing to communicate. At minimum, site visitors should be stimulated to send suggestions about what they would like to see at the site back to you via an email link. These should be taken seriously, read, and answered. Going a step further, a well-thought-out form can gather valuable marketing information. The next stage is to carry out real business transactions.

## Cleaning Up the Site

Fortunately, the problems that were described earlier are very easy to correct. Getting the job done requires directives from management that state clearly the business goals for the site. Webmasters should be rewarded for furthering those goals, not for getting meaningless "hits" or using tricky technology.

Writing clear, text-based pages is extremely simple. Lots of what-you-see-is-what-you-get tools are available. Graphics should be regarded as optional add-ons, and never should be allowed to hide information or links that can be viewed using text.

Local personnel should visit their own site often, and suggest ways to improve it. The site's home page should contain a pointer to a comments form that enables customers to have their say. This form should go to a business-oriented supervisor as well as to the webmaster.

# 3

# Getting a Dial-Up Internet Account

Anyone who will be responsible for making important business decisions about your site, and anyone who will participate in creating, supporting, or protecting your site, needs to become comfortable using the Internet and familiar with Internet resources. Fortunately, today this is very easy to do by getting an individual dial-up Internet account. In this brief chapter, we will describe criteria for choosing a dial-up Internet Service Provider (ISP) and provide a few pointers on configuring dial-up TCP/IP at your desktop system.

# Finding Dial-Up Service Providers

Dial-up accounts are available all over the United States and in major cities around the world. Note that you should get a genuine Internet dial-up which is based on TCP/IP software. Do not use the proprietary software and indirect Internet access offered by online services such as America Online. Make sure that you have a true, TCP/IP-based Internet account. Otherwise, you will not be able to perform accurate tests of Internet performance and will not be able to use several important tools.

The best way to get the information that you need in order to choose a good dial-up provider is to look it up on the Internet. This creates a chicken-and-egg problem. A good way to solve it is to get someone who has an account to help you, or to sign up with a convenient provider—possibly one that offers a free trial service.

The most complete directory of Internet Service Providers is online at *thelist.iworld.com*. At this site, you can look up providers by area code. You can click on a provider's name to visit its World Wide Web site and view information about its services and prices.

# Evaluating Dial-Up Service Providers

Dial-up services vary greatly in quality and features. In order to perform your Internet research, you will need an account that provides good, reliable performance. The paragraphs that follow describe some of the features that you will want to check.

## Modem Speed

You should obtain a modem that operates at 28.8 kilobits or more per second for your system. Dial-up modems now operate at up to 56 kilobits per second. Check that the provider supports a matching modem speed.[1]

---

[1]At the time of writing, 56-kilobit-per-second modems were not yet standardized. Be sure that there are compatible modems at both ends of the call.

## ISDN Support

If you are willing to pay more to get better performance, look for a provider that offers *Integrated Services Digital Network* (ISDN) access. But before you sign up for this digital dial-up service, check to see whether your local telephone company can install this service for you and how much it will charge for the service. Many telephone companies charge by the minute, even when the ISDN call is local. It also is a good idea to find out the expected waiting time for getting this service installed. It may take weeks or months.

## PPP

PPP stands for the *Point-to-Point Protocol,* which is the correct standards-based method of carrying Internet data across a dial-up connection. Some providers only offer older methods called SLIP or CSLIP.[2] Look for a provider that supports PPP. PPP is easier to use, often performs better, is more secure, and is more reliable.

## Costs

You need to allow plenty of time for browsing and experimentation. Look for a service that charges one flat rate for unlimited access. Hourly fees add up quickly. You do not want to pay long distance charges while connected to the Internet, so look for services that require only a local telephone call.

If you use a laptop computer and travel a lot, you might prefer a provider that has a big network with local dial-ups in all major cities. Alternatively, some providers offer 800 number service. There is an hourly fee for 800 use.

If you travel, you'd better opt for ordinary dial-up. There will be few locations that will permit you to use ISDN.

## Availability

Some providers have enrolled customers faster than they have added new equipment. You do not want to hear a busy signal for nine out of every ten of your calls.

---

[2]SLIP stands for Serial Line Interface Protocol. CSLIP stands for Compressed Serial Line Interface Protocol. CSLIP is better than SLIP, but not good enough.

To choose a provider that is not overloaded, you can ask about the ratio of users per line. Better still, some providers publish their access numbers at their Web sites. You can test availability by manually dialing numbers at various times of the day to see if they are frequently busy.

For an added fee, some providers will reserve a line that is guaranteed to be available for your use at all times. This will, of course, be much more costly than an ordinary dial-up account.

## Performance

Some small dial-up providers have only a 56-kilobit-per-second connection to the Internet. This bandwidth could be filled up by two dial-up users who are downloading big Web pages or performing file transfers. A small provider needs at least a 1.5-megabit-per-second connection in order to provide good performance to a reasonable number of concurrent users.

On the other hand, some of the large national dial-up services have enormous networks and lots of bandwidth but are greatly oversubscribed, so that users frequently get busy signals when trying to connect.

You need to look for a provider with adequate bandwidth and enough dial-up lines to provide a reasonable assurance of being able to connect.

## Support

An inexperienced user sometimes needs some help in getting going. Even experienced users will have an occasional problem. And sometimes things go wrong at the provider end; the provider equipment that is supposed to answer your dial-up call might freeze and need to be restarted by support personnel.

The ideal support service is there for you 7 days per week, 24 hours per day, and is reached via a toll-free number. You may need to compromise on some of this, but the first time that you need support, you will realize how important it is!

## ISP Application Services

Your Service Provider should give you at least one electronic mail account, forward your outgoing mail, and deliver mail to your desktop. Today, the most common Internet mail delivery mechanism is the Post Office Protocol or POP. However, if you want your mail to be stored at the Service Provider's mail server, ask about the Internet Mail Access Protocol (IMAP). Chapter 11 contains a description of IMAP's advantages.

ISPs also provide access to thousands of free news groups. Some also give you access to online versions of local and national newspapers. A few provide complete news wire services, allowing you to search for and read UP, API, and Reuters stories as they break.

# Getting an IP Address

On any system, there will be a configuration menu dealing with your IP address. This address will be assigned to you automatically when you dial up. In fact, you will receive a different address every time you dial the Service Provider. Hence, be sure that "Server assigned IP address" is selected on your configuration menu. This usually is the default.

# ISP Software for Windows 3.x

Most dial-up vendors are happy to provide you with free Internet software for Windows 3.x systems. This usually consists of:

- PPP dial-up TCP/IP software (not needed for Windows 95, NT, or Macintosh because it is built in).
- A Netscape Navigator or Microsoft Internet Explorer browser.
- An electronic mail package.

Before you load your Service Provider's TCP/IP package onto your Windows 3.1 desktop, make sure that some other TCP/IP package has not already been installed there. If it has, go on to the next section. If not, there is one chore that you should do before sliding that floppy disk into your disk drive. If you skip the steps below, you may be very unhappy some day. Backing up some key files may enable you to fix a damaged system setup.

1. Create a new directory (called INI, for example) under the root directory.
2. Copy your current AUTOEXEC.BAT and CONFIG.SYS files to the INI backup directory.
3. If your system uses any of the following, back them up too: PROTOCOL.INI, NET.CFG, or ISDN.INI.

4. Copy WIN.INI and SYSTEM.INI from your WINDOWS directory to the INI backup directory.

# Pre-existing Windows 3.x TCP/IP Software

If TCP/IP already is installed on your computer, loading a second communications package will turn your well-behaved computer into a neurotic monster. In short, don't do it!

If some unfamiliar TCP/IP package already is in your system, you may need to talk to someone on your provider's support staff who can help you configure the installed package correctly and get off to a good start. Fortunately, veteran ISP staffers have seen them all.

If you want to remove the package that is on your desktop and replace it with your ISP's software, be very careful. Erasing the directory that contains the TCP/IP software will not do the trick. The installation undoubtedly changed the files that were listed in the previous section (AUTOEXEC.BAT, CONFIG.SYS, etc.). You will need to read the product documentation and proceed carefully. You may need to call the product vendor's support staff in order to get the job done.

# Windows 95, NT, and Macintosh

Today, Windows 95, Windows NT, and Macintosh systems include TCP/IP software for both dial-up and LAN use. All that you need to do is activate it and configure it. We'll look at some sample configuration screens later in this chapter.

# Adding More Applications

Once you have installed TCP/IP on your desktop, you may wish to add one of the many application packages that are available for desktop systems. These packages offer a lot of extras. In addition to including many client applications, some provide World Wide Web and file transfer servers that you can run on your own computer.

# TCP/IP Dial-Up Configuration

You will need to obtain the following information from your service provider:

- One or more access numbers that you will dial.
- A username.
- A password.
- The addresses of two or more Domain Name Servers.
- The name of the mail server from which you will obtain your incoming mail.
- The name of the mail gateway to which you will forward your outgoing mail.[3]
- The name of your news server.

# Configuration for Windows 3.x

Installing TCP/IP on a Windows system has become very easy. There are many inexpensive and easy-to-install packages for Windows 3.1. Each product has its own configuration menus. Figure 3.1 shows two of the dial-up configuration menus for Netmanage Chameleon.

The locations of your electronic mail server and gateway and your news server are entered into separate electronic mail and news client application packages, or into browser configuration screens if you intend to use your browser as your mail and news client.

# Windows 95 Installation and Configuration

If you look up "Connecting to the Internet using Dial-Up Networking" in Windows 95 Help, you will be led through every step in the process of getting ready to connect to the Internet. We'll describe the critical steps below.

---

[3]This often is the same as the mail server used for incoming mail.

**Figure 3.1**
Two Chameleon dial-
up configuration
menus.

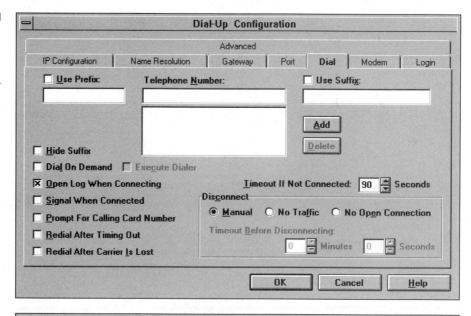

Make sure that your modem is installed. If it is an external modem, connect it, plug it in, and turn it on. Your computer needs to communicate with
the modem during the hardware install step. The help screen will point you
to the modem setup menu. Or, to check your modem directly, select:

Start/Settings/Control Panel/Modems

If your modem is not installed, click on **Add** and follow the instructions.

Then, check whether Dial-Up Networking is installed. If it is, you will see its icon after you click on the **My Computer** icon on your desktop. If not, you will have to install it using:

Start/Settings/Control Panel/Add/Remove Programs

Once Dial-Up Networking is installed, if you wish, you can use your right mouse button to drag a copy of the icon to the **Start** menu.

Next, check whether TCP/IP is installed. To do this, select:

Start/Settings/Control Panel/Network

If TCP/IP is not installed, click on:

Add, Protocol, Add, Microsoft, TCP/IP, OK

When you click on **TCP/IP -> Dial-Up Adapter, Properties**, you will see configuration menus for Windows 95 TCP/IP, as shown in Figure 3.2. The normal situation is that you will obtain an IP address automatically. You should disable DNS and enter DNS server addresses on the individual dial-up menus that will be created in the next step.

Finally, open up **Dial-Up Networking** and create separate icons for each of your dial-up destinations. To do this:

- Double-click **Make New Connection.**
- Enter a name for the connection and click **Next.**
- Enter the dial-up telephone number, click **Next** and **Finish.**

You might think that you are finished, but actually you are just beginning. To configure this dial-up destination, right click on the new icon and select **Properties.** The screen in Figure 3.3 is displayed. You can click the **Configure** button to check out your modem properties. Then go back and click the **Server Type** button to configure TCP/IP. Figure 3.4 displays the screen that appears.

Check the top field to make sure that PPP has been selected, and uncheck NetBEUI and IPX/SPX so that TCP/IP is the only protocol that is selected.

Next click **TCP/IP Settings** and enter the addresses of your ISP's Domain Name Servers. Then click **OK** until you have exited the configuration menus. To dial this destination, just double-click on its icon and enter your user name and password. There is a check box that allows you to save your password.[4]

---

[4]The password will not be stored permanently until you have made a successful connection.

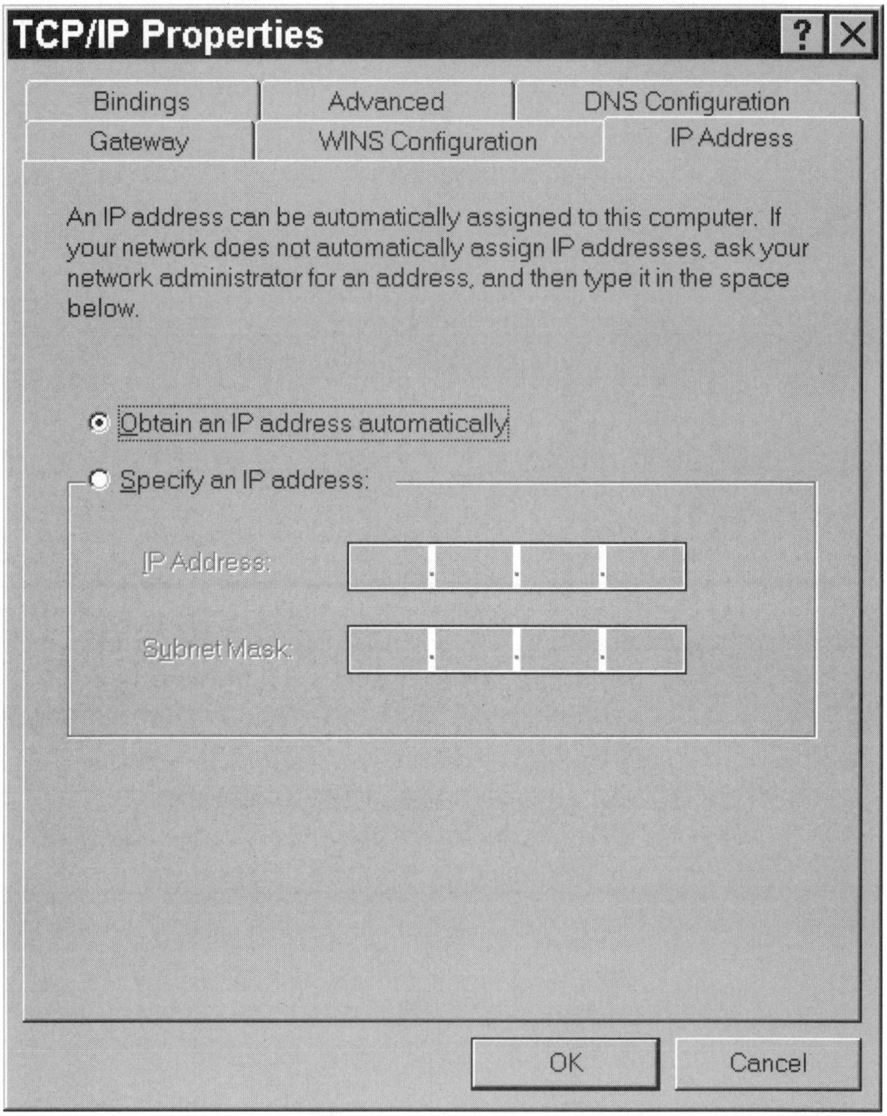

**Figure 3.2**
Configuring TCP/IP
Properties for
Windows 95.

**TCP/IP Properties**                                    **?** **X**

| Bindings | Advanced | DNS Configuration |
| Gateway | WINS Configuration | IP Address |

An IP address can be automatically assigned to this computer. If
your network does not automatically assign IP addresses, ask your
network administrator for an address, and then type it in the space
below.

⦿ Obtain an IP address automatically

○ Specify an IP address:

   IP Address:          [    .    .    .    ]

   Subnet Mask:         [    .    .    .    ]

                                    OK          Cancel

# Windows NT Server Dial-Up Installation and Configuration

Configuring NT Server dial-up for version 4 is very confusing. Fortunately,
it is something that you only need to do once. You may wish to look

**Figure 3.3**
Properties for a new
dial-up location.

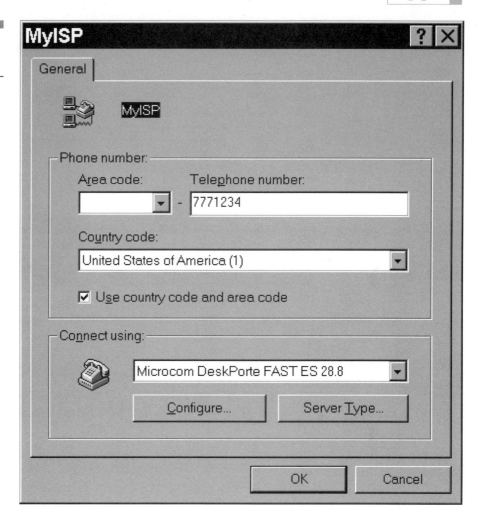

at the steps under the Help index topic Dial-Up Networking: Internet
connections.

As was the case for Windows 95, the first step is to install your modem. If
it is an external modem, connect it, plug it in, and turn it on. You can check
up on your modem directly by selecting:

Start/Settings/Control Panel/Modems

If your modem is not there, then add it.

Next, go to Network Properties by selecting:

Start/Settings/Control Panel/Network

**Figure 3.4**
The Windows 95
TCP/IP Server Types
configuration screen.

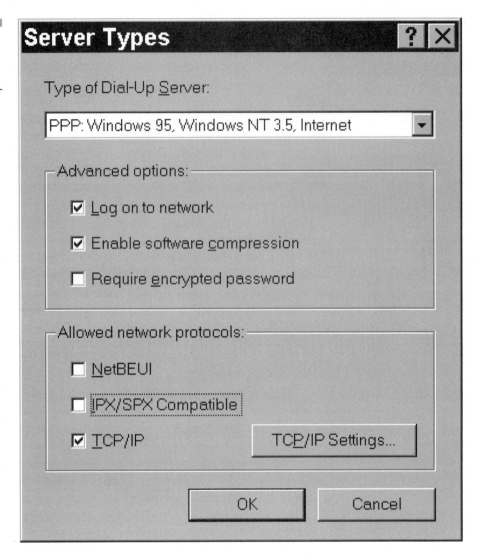

**Figure 3.4**
The Windows 95
TCP/IP Server Types
configuration screen.

If the TCP/IP protocol is not listed, then add it. The resulting screen should look like Figure 3.5. Next you have to make sure that Dial-Up Networking is installed. Look for:

Start/Accessories/Dial-Up Networking

On Windows NT Server, this is the client portion of the Remote Access Service (RAS). If Dial-Up Networking is not there, then Step 5 under the Help topic, "Install Dial-Up Networking," gets the Dial-Up Networking software

**Figure 3.5**
Windows NT Server
Network Protocols
screen.

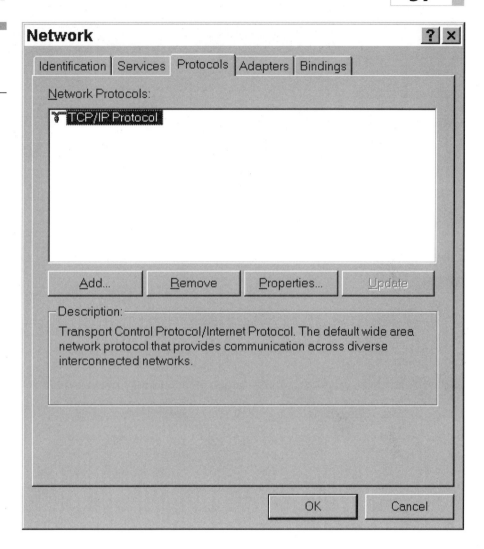

into your system very easily. Figure 3.6 shows the main dial-up configuration
screen.

From this screen, you can add new dial-up destinations, click the More
button to edit entry and modem properties (as shown in Figure 3.7), and click
the Dial button to connect to a remote site.

Three of the tabs on the menu in Figure 3.7 are important. Choose the Basic tab and the Configure button to configure your modem.

Choose the Server tab to make sure that PPP, TCP/IP, Enable Software
Compression, and Enable PPP LCP Extensions are selected, as shown in

**Figure 3.6**
The Main NT Server
Dial-Up Networking
menu.

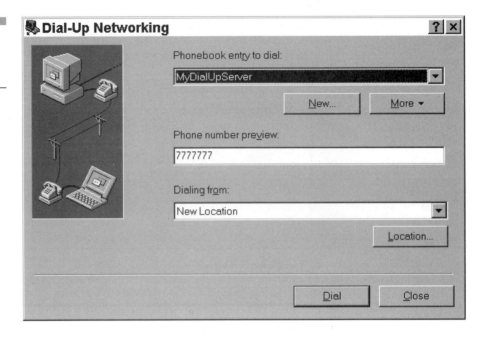

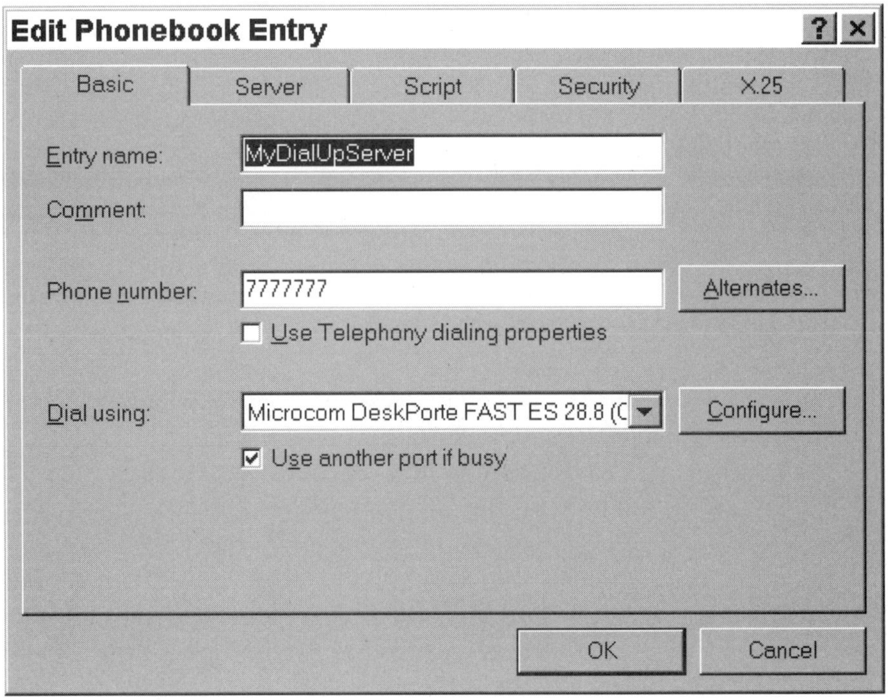

**Figure 3.7**
The NT Edit Phone-
book Entry screen.

**Figure 3.8**
Configuring NT
Dial-Up TCP/IP on
the Server menu.

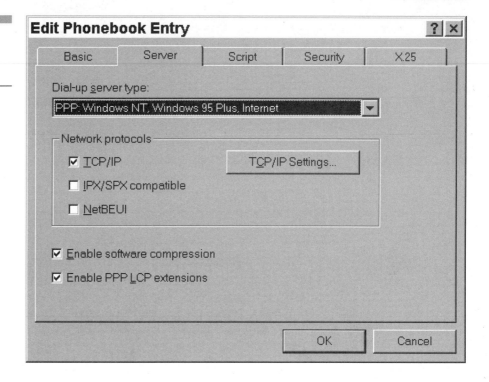

Figure 3.8. Click on the **TCP/IP Settings** button to enter the IP addresses of your Service Provider's Domain Name Servers.

Finally, take note of the **Security** tab. When you click **Dial** for this entry, you will be prompted for your username and password. There is a check box that allows you to save your password so you do not have to enter it again. In this case, the password box will not be present the next time you start the Dial-Up Networking program. But suppose that you have to change your password? You will have to enter the **Security** screen and click the **Unsave Password** button. When you click **Dial** again, the password box will have reappeared.

# 4

# Name Registration

Chapter 1 included a very brief description of Internet computer names, which are called domain names. Before you can connect to the Internet, you must choose and register a domain name. Registering a domain name is not just the simple process of reserving a name, such as *xyz.com*. You will not be allowed to register until you are ready to participate in the Domain Name System (DNS), the Internet's distributed directory lookup service. The whole procedure can take a while, so this is something that needs to be worked on as soon as possible. This chapter provides the information that you need in order to understand and participate in the registration process and to keep your site's registration up-to-date when changes must be made.

While this book was being written, a plan was being put in place to overhaul the entire Internet registration system. We will describe the current system, and also indicate some of the changes that are contemplated. The basic series of actions you must take to join the Domain Name System and get registered are unlikely to differ very much, even after modifications to the registration system have been carried out.

# Overview of the Process

The easiest way to get registered is to delegate most of the work to an Internet Service Provider (ISP). If you have not yet chosen the provider who will connect you to the Internet, you may be able to find a provider who will help you to register for a nominal fee. The series of steps that you need to carry out is:

1. Decide where your organization best fits into the Internet's naming scheme (*com, org,* etc.).

2. Find a name that no one else is using.

3. Obtain a domain registration form from an appropriate registrar. Your provider can do this, but you should examine a copy of the form.

4. Choose someone in your organization to act as the administrative point of contact. Make sure that this person's name appears on the form that is submitted. Or, if you wish more flexibility, identify the person by job role instead of by name.

You might think that you would now be ready to file your form with the Internet registration service. However, Internet registration is not about reserving names. The registration service is tied into managing the Internet Domain Name System directory service. Recall that in Chapter 1 we introduced the Domain Name System, which is a set of automatic directories used to translate between computer names and addresses.[1] These directories are called Domain Name Servers, or just "name servers" for short.

The Internet registration service requires every site that connects to the Internet to do its share of work for the directory system. This means that before your site can exchange data with the Internet, there must be at least two directory servers that are up and ready to answer queries relating to your domain name. You will not be allowed to register your domain name until two servers are operating and responding to queries about your selected domain name.

Fortunately, even if you are not yet really connected to the Internet, your Service Provider can help you to qualify for registration by entering your selected domain name into a couple of its own directory databases.

Now that we understand the overall process, we are ready to look at the details.

---

[1] Later, we will see that there is a lot of other useful information stored in DNS directories.

# Structure of Internet Computer Names

Internet computer names are organized into a big tree,[2] as shown in Figure 4.1. Names enclosed by dotted boxes are proposed additions to the tree.

Names are made up of a series of labels separated by periods. A label can contain letters, numbers, and the minus symbol ("−"). The case of letters does not matter; that is, *abc.com, ABC.COM,* and *Abc.Com* all are treated as the same name. There are very few rules:

- The first character in a label must be a letter and the last character cannot be a minus.

- Each label can be up to 63 characters long.

- The total length of a name can be far longer than any name that a sensible person would ever choose (over 240 characters).

The last (rightmost) label in your computer names will depend on your type of organization or the country in which you are located. This final label is called the *top level domain name.* The combination of the final two labels—*mcdonalds.com, red-cross.org, yale.edu*—is called the *second level domain name.*

The classical top level domain names are listed in Table 4.1. The first three—*com, org,* and *net*—are called generic. This means that they are not tied to a specific country. At the time of writing, seven new generic top level names had been proposed. These are listed in Table 4.2. Additional generic names may be added in the future.

---

[2]Computer users are familiar with trees that are upside down, like this one. Other people might prefer to view this as an organizational chart.

**Figure 4.1**
The naming tree.

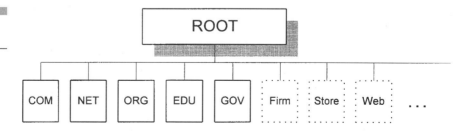

**TABLE 4.1**

Classical Types of
Internet Names

| Name | Description |
|------|-------------|
| Com | Used for commercial, for-profit organizations. |
| Org | Miscellaneous. Usually for non-profit organizations. |
| Net | Originally intended for network infrastructure machines and organizations, but now also used by businesses. |
| Edu | U.S. 4-year degree granting institutions. |
| Gov | U.S. federal government agencies. |
| Mil | U.S. military organizations (army, navy, etc.). |
| US | U.S. state and local government agencies, schools, libraries, museums, and other miscellaneous organizations. |
| Int | Organizations (such as NATO) formed by international agreement. |
| UK, DE, etc. | Naming subtrees for countries other than the United States. Each country can choose its own lower-level naming structure. |

# Name Administration

Somebody has to make sure that organizations that connect to the Internet choose and use different names and publish these names at Domain Name Servers. This supervision is done by a variety of registration organizations that are responsible for different parts of the naming tree. Some of these supervisory organizations also participate in address registration. In this chapter we

**TABLE 4.2**

New Types of
Generic Internet
Names

| Name | Description |
|------|-------------|
| Firm | For businesses or firms. |
| Store | For businesses offering goods to purchase. |
| Web | For entities emphasizing activities related to the World Wide Web. |
| Arts | For entities emphasizing cultural and entertainment activities. |
| Rec | For entities emphasizing recreation and entertainment activities. |
| Info | For entities providing information services. |
| Nom | For those wishing to use individual or personal naming. |

will concentrate on name registration. Chapter 7 will discuss address registration issues.

The largest share of domain name registrations currently is carried out by the InterNIC Registration Service. The InterNIC has been responsible for the most popular parts of the naming tree. We'll discuss the InterNIC and several other registration organizations in the sections that follow.

# The InterNIC
# Registration Service

In 1993, primary administration for several top level domain names was placed in the hands of the Registration InterNIC, which was awarded a grant by the National Science Foundation. The Registration InterNIC has had responsibility for names that end in *com, net, org, edu,* and *gov.* All second level domain names ending in these labels are entered into files at the Registration InterNIC site.

The Registration InterNIC also has been responsible for adding a new label to the top of the naming tree whenever an additional country connected to the Internet.

The Registration InterNIC maintains a database (called the *Whois* database) that lists Domain Name System administrators along with their email addresses and telephone numbers. This is a very important resource. If there is a problem with a site's directory, it is important to get in touch with a responsible party as soon as possible. More importantly, this is the person to call if there are connectivity or security problems related to the site. Later in this chapter, we'll look at some dialogs that demonstrate how you query this database using a *Whois* client application. The last section in this chapter contains more information about *Whois* client syntax, along with additional examples.

You can get more information about the InterNIC registration service by connecting to the InterNIC web site:

*http://www.internic.net/*

Click on Registration Services. Later in this chapter, we will describe their registration procedures in detail.

The Registration InterNIC is operated by Network Solutions, Inc., in Herndon, Virginia. Its current National Science Foundation contract expires in January of 1998.

# Updated Administrative Structure

The internationalization of the Internet, competition for second level names under the *com* domain, and trademark issues made it clear that it would be beneficial to restructure the way that name assignments are administered.

In the future, there will be many registrars at locations all around the world. Initially, these organizations will handle registrations under the new top-level generic domain names. Eventually, they will share responsibility for the generic *com*, *org*, and *net* top-level domains currently managed exclusively by the Registration InterNIC.[3]

# The *US* Domain

Registration for the *US* domain is coordinated by the Internet Assigned Numbers Authority (IANA) at the University of Southern California Information Sciences Institute (ISI). You can get information about registering under the *US* domain or access a registration template at their web site, *http://www.isi.edu:80/in-notes/usdnr/* or at their file transfer site, *ftp://ftp.isi.edu/*.

There are many categories under the *US* domain, including:

- State, county, and local government
- Some federal government agencies
- Public and private K–12 schools
- Libraries
- Community colleges and technical schools

Names under the *US* tree follow well-defined patterns. Sample patterns include[4]:

&lt;school-name&gt;.*k12*.&lt;state&gt;.*us*

&lt;school-name&gt;.*pvt.k12*.&lt;state&gt;.*us*

---

[3]Overall supervision of name registration is being shifted to the DNS Policy Oversight Committee, which is made up of representatives from the Internet Society, the Internet Assigned Numbers Authority, the Internet Architecture Board, the International Telecommunications Union, the International Trademark Association, the World Intellectual Property Organization, and an international Council of Registrars. Connect to *http://www.isoc.org/* for pointers to information on new registration policies.

[4]See *ftp://ftp.internic.net/rfc/rfc1480.txt* for details.

<lib-name>.*lib*.<state>.*us*

<org-name>.*state*.<state>.*us*

Like the Registration InterNIC, ISI also maintains a database of second-level domains. However, the second-level domains under *US* do not correspond to companies or organizations. The ISI database entries point to other name servers. For example, there are pointers to servers for state domains such as *al.us* or *tx.us*. The actual work of registering end-user organizations is delegated to many state and regional registrars.

## The *Int* Domain

The Internet Assigned Numbers Authority at ISI also is in charge of names in the *int* domain. This means that they handle registration for organizations (such as NATO) that have been created via international treaty.

## The *Mil* Domain

The Department of Defense Network Information Center is responsible for administering military domain names. Registration forms and general information are available at the web site:

*http://nic.ddn.mil/*

Their registration center is currently located at:

7990 Boeing Court

M/S CV-50

Vienna, VA 22183-7000

## Finding Registries Outside of the United States

There are many regional and national Network Information Centers that have been set up to register names under country codes such as *uk, ch,* or *de*. There is a very good list of registries at the Yahoo search site at:

*http://www.yahoo.com/Computers_and_Internet/Internet/*

*Domain_Registration/Network_Information_Centers/*

For example, this Yahoo page has a pointer to *http://www.nic.fr/*, a site that maintains the list of domains under the DNS top level domain for France (*fr*).

Each European or Asian-Pacific country can define the structure of its own naming subtree, set its own policies, and provide its own registration forms and fees. The two sections that follow provide brief descriptions of two important regional Network Information Centers.

# RIPE European Center

Reseaux IP Europeens (RIPE), located in Amsterdam, coordinates the Internet in Europe. Information is available at:

*http://www.ripe.net/*

Currently, RIPE does not register domain names. Instead, it provides pointers to national registries. The site maintains a *Whois* database identifying the domain name administrators for European countries. RIPE also provides pointers to many European Internet Service Providers.

RIPE is in charge of IP address assignments in Europe. It does this by delegating chunks of addresses to European ISPs. An end-user organization must submit its address request to an ISP, which will either provide addresses out of its own space or will submit the request to RIPE.

# Asian-Pacific Network Information Center

The Asian-Pacific Network Information Center (APNIC), located in Tokyo, coordinates the Internet for Asia and the Pacific Rim countries. Information is available at:

*http://www.apnic.net/*

Like RIPE, APNIC does not currently register names, but it does provide pointers to national registries. It also provides access to a *Whois* database identifying the site administrators for Asian-Pacific countries, and pointers to Service Providers that operate in the region. APNIC is in charge of IP address assignments in the Asian-Pacific region.

## *RWhois*

There are *Whois* databases operated by many different registration entities and scattered around the world. Currently, a distributed *RWhois* service is being set up on an experimental basis. This service will allow you to send any query to any *Whois* server and get a response that points you closer to a server that can answer the query.

# Choosing Your Domain Name

If you have not already chosen your own Domain Name, now is the time to nail one down. Between January of 1995 and July of 1996, the number of registered domain names increased from 71,000 to 488,000. By September of 1996, the number was up to almost 639,000, and registrations have continued their upward climb since then.

Start by choosing a few names that would be suitable for your organization. Then check to see if any of them are available. A quick way to find out whether a name already has been taken is to use the *nslookup* application. The *nslookup* application lets you talk directly to a name server.

There are lots of different implementations of *nslookup*. My favorite version was written by a British company called Ashmount Software and runs on Windows 3.1, Windows 95, or Windows NT. It has an elegant, simple user interface and, best of all, it is free. I picked up my copy from:

*ftp://ftp.demon.co.uk/pub/trumphurst/nslookup/*

In Figure 4.2, we check whether the name *abc.com* is available. To create the query, we fill in the three boxes at the bottom as follows:

**NameServer field:** Enter the address of a nearby name server,[5] enclosed in brackets; in this case, [130.132.1.9].

**Name field:** Enter the domain name that you want to check: in this case, *abc.com*.

**Type field:** Use the arrow to display the list of query types, and choose type *Name Server*.

---

[5]For example, a name server operated by your ISP.

**Figure 4.2**
Using Ashmount's
*nslookup* to check a
domain name.

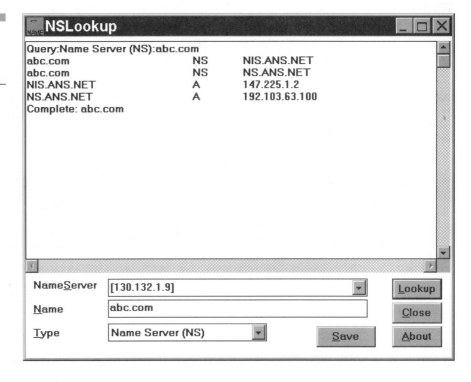

If you wanted to use the name *abc.com*, you are out of luck. Somebody else got there first. When a name already is taken, there are two possibilities:

1. Another organization may legitimately have the same title as yours and the name may be in real use.

2. Someone may have obtained the name on speculation, hoping to sell it later.

Network Solutions, which currently operates the Registration InterNIC, does not have the resources or the desire to monitor whether an applicant is the best candidate to use a particular name. They are kept quite busy simply maintaining the database of domain names, administrators, and servers. According to their current policy statement:

- They register names on a "first-come, first-served" basis.

- There is a clause on the registration form stating that the requester certifies that, to her/his knowledge, the use of this name does not violate trademark or other statutes. However, Network Solutions does not check up on trademarks or arbitrate disputes.

- Network Solutions leaves the settlement of disputes over names to external legal procedures.

Legal processes can be slow and costly. There have been several cases in which brand names were bought back from a registrant.[6] If you do not want to fight or pay for the right to use *abc.com*, you could try a similar title, such as *abc-corp.com*. We will check this one out using the classic implementation of *nslookup* at a Unix computer. This text-based *nslookup* also runs on NT systems, where you can start it in a command window. (Just choose Start/Programs/Command Prompt.)

```
> nslookup
Default Server:  DEPT-GW.CS.YALE.EDU
Address:  128.36.0.36

> set type=ns
> abc-corp.com.
Server:  DEPT-GW.CS.YALE.EDU
Address:  128.36.0.36

*** DEPT-GW.CS.YALE.EDU can't find abc-corp.com.: Non-existent
host/domain
```

This time you are in luck. You could use *abc-corp.com* as your name.

The *nslookup* tool will perform lookups for any name, including those that end with a country code such as *uk*. An alternative way to search for free names is to use the *Whois* menu at the Registration InterNIC Web site:

*http://rs.internic.net/cgi-bin/whois*

However, queries entered into this menu cannot be used to search for names outside of the InterNIC's part of the naming tree.[7]

# Checking Names by Using Zone Files

If you have a problem finding a usable name, you might get tired of looking up one name at a time. All of the names for *com, org, net, edu,* and *gov* are stored in "zone" files that you can copy from:

*ftp://rs.internic.net/domain/*

---

[6]There is a plan to place the Arbitration and Mediation Center of the World Intellectual Property Organization in charge of setting up administrative domain name challenge panels. An organization will be able to challenge the use of a name for which is has demonstrable intellectual property rights.

[7]They have an *RWhois* menu that you could try.

The zone files have names like *com.zone.gz* and are compressed via the free *gnu zip* program.[8] But watch out! Even in compressed format, the *com.zone.gz* file is very large. And when you copy the file, remember that a compressed file has a binary format.

Here are a few records from the *com.zone* directory file. They identify the names and addresses of two name servers that provide directory service for names that end in *NBC.COM*. (The number in the second column is an optional parameter that we will explain in Chapter 12.)

```
NBC.COM.            172800      NS      CRDRAS.GE.COM.
CRDRAS.GE.COM.      172800      A       192.35.44.7
NBC.COM.            172800      NS      NS.GE.COM.
NS.GE.COM.          172800      A       192.35.39.24
```

Note that neither of the directories for *NBC.COM* is inside of the NBC network. The directories are maintained by their ISP, which is perfectly acceptable.

# Setting Up Domain Name Servers

We have urged you to register your domain name immediately—possibly before you even start to plan your site. But this gives you a chicken-and-egg problem. The InterNIC rule is:

"At the time of the initial submission to Network Solutions of the Domain Name request, the Registrant is required to have operational name service from at least two operational domain name servers for that Domain Name. Each domain name server must be fully connected to the Internet and capable of receiving queries under that Domain Name and responding thereto. Failure to maintain two active domain name servers may result in the revocation of the Domain Name registration."

The InterNIC requirement actually is not very difficult to satisfy. The only query the servers need to respond to is, "Please return the names and addresses of the Domain Name Servers for *abc-corp.com.*"

Fortunately, many Internet Service Providers are very happy to solve this problem for you. Keep in mind that your Domain Name Servers do not have to be located at your site or run by you. ISPs simply add some skeleton data-

---

[8]A shareware program that can uncompress *gz* files is available at *http://www.winzip.com/*.

base entries to one of their own Domain Name Servers. Entries like the following would suffice:

```
ABC-CORP.COM.          NS      NS1.BIGISP.NET
NS1.BIGISP.NET.        A       130.15.1.1
ABC-CORP.COM.          NS      NS2.BIGISP.NET
NS2.BIGISP.NET.        A       130.15.1.4
```

These records identify *NS1.BIGISP.NET* and *NS2.BIGISP.NET* as the current official Domain Name Servers for *abc-corp.com*, and provide the addresses for these servers. Your name servers do not have to contain records that point to machines on your own network. Initially, all they need to do is identify their own locations.

Later on, you can add entries for one or more other computers, such as a mail server or a World Wide Web server. Most ISPs are happy to continue to run domain name directory services for you for a nominal fee.

As soon as the Domain Name Servers are ready, you (or your ISP) should fill in a form that requests the domain name that you have chosen and identifies the names and addresses of the Domain Name Servers. We will describe that form in detail a little later. The InterNIC will add your name server entries to an appropriate zone file (*com.zone* in the case of *abc-corp.com*).

## Root Database

The contents of the zone files are made available to the world via a *root* directory database, which is replicated at several sites. This database is what ties the entire Domain Name System together. Computers around the world will find your name servers by looking up your name via the root directory. We'll show step-by-step lookups in Chapter 12. Currently, the InterNIC Registration Service administers the root directory.

## Registering Your Domain Name

Although your ISPs will be happy to register on your behalf, it is a good idea for you to participate in the process. You should review the form before it is submitted and ask to be notified of the tracking number that is assigned at submission time. If you don't participate and there are any glitches, you may be powerless to fix whatever problem may have arisen.

At the time of writing, the registration procedure was simple. It could even be done by filling in a form at the Registration InterNIC World Wide Web site. However, it is a better idea for the person in charge of filing to retrieve a copy of the registration template, fill it in, print it, and review it. The completed form must be submitted via electronic mail.

The same form will be used later to report updates such as adding a new Domain Name Server or changing contact personnel.

## Registering via Email

The steps to follow to register via email are:

1. Obtain an electronic copy of the domain registration form and fill in the requested information. A form can be retrieved via file transfer from:

   *ftp://rs.internic.net/templates/domain-template.txt*

2. Email the form to *HOSTMASTER@INTERNIC.NET*. An auto-reply with a tracking number will be sent back. You can check up on the status of your registration by entering this number on a tracking screen at the *rs.internic.net* Web site.

3. The filer will be notified via email when the registration has been completed. You can check that your information has been entered correctly by performing an *nslookup* query and a *Whois* query.

4. The billing contact identified on the form will be billed $100. This fee will cover two years of registration.

The InterNIC updates its procedures from time to time. It is a good idea to check the InterNIC Registration World Wide Web site for current information on registration policies and procedures.

## Domain-Template Form

The domain registration form is not complicated. It asks:

■ Whether this is a new registration, a modification, or a deletion.

■ What type of organization you have and why you are registering.

■ Your selected domain name.

■ The name and address of your organization.

- Administrative contact: the person (or role) who can speak on behalf of the organization and answer questions about the organization's planned naming structure.

- Technical contact: the person (or role) responsible for running and maintaining the servers and their database files. If technicians elsewhere on the Internet detect a problem with your server or with your information, this is the person to be notified.

- Billing contact: the person (or role) responsible for paying the bill. There is an initial $100 fee that covers the first two years of registration. Then a $50 per year maintenance fee is charged.[9]

- The name and address of your primary name server. (This is where data is entered.)

- The name(s) and address(es) of your secondary name server(s). (These are copies of the database.)

The name of each contact person must be entered in the form Lastname, Firstname Middlename. Note that contacts can be identified by role, so that if a new person is assigned responsibility for one of the functions, the InterNIC will not have to be notified.

If this information is entered at the Web site, the completed information will be mailed back to you, and then you must remail it back to the InterNIC. This provides a minimal validation of your identity; at the least, you have the email address you claim to have.

# InterNIC *Whois* Database

Information about your site and your contact personnel is entered into the *Whois* database. The *Whois* database is an important resource for technical support personnel. Sometimes the only way to resolve persistent problems in reaching a remote site is to get in touch with an administrator at that site.

There are several ways that users can access *Whois* databases. Unix systems provide a simple text-based user interface. The detailed syntax of the Unix *Whois* command can be a nuisance to learn. Some TCP/IP application packages include a graphical user interface that runs right on the desktop. Or, you might prefer to use the query form at the InterNIC World Wide Web server that we mentioned earlier (*http://rs.internic.net/cgi-bin/whois*).

---

[9]The new registration services may establish other billing policies.

Here is the result of a Unix *Whois* query on the name *novell.com*. Recall that there are a lot of *Whois* databases spread all over the world. In the query, we use the host ("-h") parameter and identify the name of the host at which the InterNIC database is located (*rs.internic.net*).[10] Note that one individual is acting as Novell's administrative, technical, and billing contact.

```
> whois -h rs.internic.net novell.com


Novell, Inc. (NOVELL-DOM)
   122 East 1700 South
   Provo, UT 84606
   USA


   Domain Name: NOVELL.COM


   Administrative Contact, Technical Contact,
      Zone Contact, Billing Contact:
      Lodge, Michael  (ML205)  . . .


   Record last updated on 21-Nov-95.
   Record created on 20-Nov-89.


   Domain servers in listed order:


   NS.NOVELL.COM       137.65.1.1
   NS.UTAH.EDU         128.110.124.120
   NS1.WESTNET.NET     128.138.213.13
```

Note the phrase NOVELL-DOM at the top of the entry. This is a string called a *handle* that the InterNIC has assigned to the database entry so that the entry will have a unique index.

Contact personnel can be looked up via a direct query of the InterNIC *Whois* database. Since a person's name (such as John Smith) may not be unique, a handle index also is assigned to each contact person entry. The handle consists of the person's initials followed by a number that makes the index unique.

Here is a response to a query on the name Joseph Paolillo, which is entered inside quote marks below. The quotes are needed because of the space between the last and first names. The query will work without the quotes if the space is left out. Note the handle index, JP218, which appears in parentheses after the name in the response.

---

[10]If we want to make queries about networks in Europe or the Pacific Rim, we would need to send queries to *whois.ripe.net* or *whois.ripe.apnic*.

```
> whois -h rs.internic.net "Paolillo, Joseph"
Paolillo, Joseph (JP218)                    joseph.paolillo@YALE.EDU
   Yale University
   175 Whitney Avenue
   New Haven, CT 06520
   ( 203 ) 432.6673
   Record last updated on 31-May-96.
```

# ISPs and Registration

As long as your organization is identified on the form, you will own the domain name even if your Service Provider registers for you. However, it is important to assure that one of your own internal people is listed as the administrative point of contact. Otherwise, you really will not have control over your entry.

For example, if you should decide to switch Service Providers, you will discover that you are totally dependent on your Service Provider to make the required change request to the registration database. If you have had problems and are about to dump the provider, you may discover that you are in for a long delay.

# Non-Standard Registrations

There may be dozens of businesses that justifiably would like to be registered as *mcdonalds.com,* but only one can succeed in doing so. During the past few years, many people have felt that an expansion in the number of top level names was needed in order to reduce name collisions. As we have seen earlier, official action has been taken to bring this about.

Some Internetters did not wait for this action. They became irritated with the slow pace of the standards process in meeting their needs. Some also felt that the InterNIC was not justified in charging a registration fee. On the Internet, it does not take long for action to follow irritation. They formed alternative unofficial registration organizations[11] that are registering names into their own naming trees.

---

[11]For example, see *http://www.alternic.net/.*

# Updating Your Registration

Over time, it is likely that you might change the location of one of your name servers or, if you get a lot of queries, you might add a new name server. The same form that is used for registration also is used to mail in a request that updates your contact information or changes your Service Provider (or even deletes your registration entry).

Many organizations simply send their updates to the InterNIC in an unsecured email message. For years, the InterNIC conducted business by simply checking the From: field in an incoming email message to see if it came from the email address of a registered administrative or technical contact. *Unfortunately, it is very easy to forge an email message and make it appear that it came from anyone at all.*

This fact was brought to everyone's attention in a dramatic fashion. On Christmas Day in 1994, a hacker named Kevin Mitnick broke into the San Diego Supercomputer Center. Among other things, Mitnick rifled files at a computer used by security specialist Tsutomu Shimomura. Shimomura pursued and tracked down Mitnick, and later coauthored a book called *Takedown* that described the chase. He marketed the book at a site called *www.takedown.com.*

An anonymous hacker cut off access to Shimomura's site by forging email messages to the InterNIC. This hacker removed Shimomura's domain name and replaced it with *takendown.com.* The moment that the registration database was changed, Shimomura's site became totally unreachable by anyone who did not happen to know its IP address.

According to current policy, the InterNIC will not accept requests to change a domain name. However, it is disquieting to think that an email request to delete your entire registration might be honored. Fortunately, you can protect your entry if you are willing to invest a small amount of effort.

# Improving Contact Security

Specified people called *contacts* are responsible for all of the information that the InterNIC maintains about your resources. You list contacts on your domain form and on additional host forms that provide information about the systems that act as your Domain Name Servers.

To improve the security of your interactions with the InterNIC, the first step is to have each contact person fill out and submit a Contact form. The Contact Template can be retrieved from:

*ftp://rs.internic.net/templates/contact-template.txt*

On this form, a contact person can specify how two important functions, notification and authentication, should be carried out.

## Notification

When you register as a contact, you can ask to be notified (via electronic mail) after any change is made to items for which you are responsible. Or, you can choose a safer alternative, and ask to be notified *before* any change is made to such items. The change will not be carried out until you send an acknowledgment.

## Authentication

The default method that the InterNIC uses to validate an update request (called MAIL-FROM validation) is a simple and insecure check of the From: field in an incoming electronic mail message. Since mail forgery is all too simple, this provides no protection.

You can improve security slightly by choosing an option called CRYPT-PW. On the Contact Template form, the contact:

- Chooses a password.
- Encrypts the password using the Unix *crypt* function.
- Writes the encrypted version on the registration form.
- When an update is requested, the cleartext form of the password must be included with the request.

This method is not very safe because hackers occasionally succeed in eavesdropping on Internet email. A hacker who has intercepted a message containing a cleartext password would be able to forge change requests.

The PGP (Pretty Good Privacy) authentication option is the only one that provides real security. PGP is security software that is based on public/private key pairs and the use of digital signatures. These security mechanisms are explained in Chapter 10.

To use PGP, a contact must do the following:

1. Download PGP software. Both freeware and commercial software are available at *http://www.pgp.com/*.

2. Generate a public/private encryption key pair using this software.

3. Use a Contact Template form to indicate that PGP will be used.

4. Register the public key with the InterNIC.

5. Include a PGP digital signature in all future update requests.

The contact's public key is registered by sending an electronic mail message with the format:

```
To: pgpreg@internic.net
From: johndoe@example.com
Subject: add

---BEGIN PGP PUBLIC KEY BLOCK---
Version: 2.6
>Public key goes here>
---END PGP PUBLIC KEY BLOCK---
```

To get more details, send the following message:

```
To: pgpreg@internic.net
From: johndoe@example.com
Subject: help
```

A contact that is protected by authentication information is called a *guardian*. Details of the InterNIC guardian system are available at the InterNIC's Web site. This system was worked out before commercial secure email products based on a market standard called S/MIME were available. Now that such products have entered the marketplace, perhaps the system will be updated to support these products, which will be widely available and easy for the contacts to use.

Alternatively, registration sites may enable contacts to use certified electronic signatures to support secure interactive updates at their Web servers. Chapter 10 describes electronic signatures and certification.

## Host Changes

To add a new Domain Name Server or change existing host records, an authorized contact must submit a Host Template form. The Host Template is available at:

*ftp://rs.internic.net/templates/host-template*

# More About *Whois*

*Whois* is a very useful tool, so it is worth a second look. (Check the documentation of your own *Whois* client program for more details.)

*Whois* will search for anything in its database that matches a string that you give it. For example, here are a few of the responses to a query for "yale." (The names of people have been changed.)

```
> whois -h rs.internic.net yale
. . .
Harrison, Yale (YH1021) yale@XYZ-CORP.COM  444-888-4945
. . .
YALE LA FONTE FECHADURAS S/A. (LAFONTE-DOM) LAFONTE.COM
Yale Capital Group, Ltd.(YALECAPITAL-DOM)YALECAPITAL.COM
Yale Club of Cyberspace (YALE3-DOM)       YALE.ORG
```

You can search for IP addresses as well as for names. The query below finds the owner of a network address. We will be using queries of this type in Chapter 7.

```
> whois -h rs.internic.net 130.132.0.0
Yale University (NET-YALE-SPINE)
    New Haven, CT 06520

    Netname: YALE-SPINE
    Netnumber: 130.132.0.0

Coordinator:
   . . .
Alternate Contact:
   . . .
```

A general search will try to match up a query string with anything: names, addresses, handles, and electronic mail identifiers. You can use some special query formats to restrict your search, as shown in Table 4.3.

If you are entering your queries via a Unix text-based user interface, then a handle query would actually have the form:

```
> whois -h rs.internic.net \!YALE-DOM
```

**TABLE 4.3**

Special *Whois*
Query Formats

| | |
|---|---|
| Start with a period to restrict the search to names only. | .yale |
| Add an @ character to look for email identifiers. | Jones@ |
| Start with an exclamation point to restrict the search to handles only. | !YALE-DOM |

This is because the ! character is a special symbol for the Unix user interface. The preceding \ symbol says "treat the character that follows like an ordinary character instead of as a special symbol."

You can make your search even more general and look for all entries that start with a given string by adding a dot at the end. "Robert." will return matches on Robert, Roberts, Robertson, and so forth.

# REFERENCES

RFC 1480 *The US Domain*, A. Cooper, J. Postel.

Pointers to InterNIC Registration Frequently Asked Questions documents can be found in:

*http://rs.internic.net/help/index.html*

# **5**

# Deciding How to Connect Your LAN

When you are ready to connect your site to the Internet or upgrade your current Internet link, you will want to choose the best connectivity method. Many Internet Service Providers offer a very broad range of transmission choices: dedicated dial-up, Integrated Services Digital Network (ISDN), frame relay, leased lines, and others. We will introduce and describe these choices in this chapter.

In addition to choosing the type of transmission method, you need to decide whether to connect to the Internet via a router, a smart firewall router, or an application proxy firewall.

**Figure 5.1**
Connecting directly
via a router.

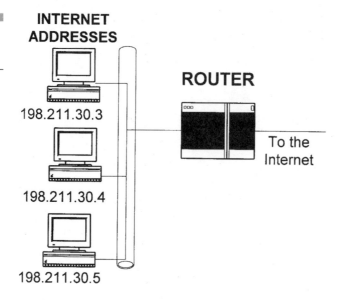

## Connecting a LAN via a Router

As we already have seen, TCP/IP networks are connected together by routers, which move traffic to its destination. If you want to route traffic between a LAN and the Internet, then you need to obtain unique public IP addresses from your ISP and configure your LAN systems with these addresses, as shown in Figure 5.1.

The router should be set up so that it examines incoming and outgoing traffic and discards dangerous traffic. This is called *filtering*. We'll provide more information about filtering in Chapter 9, which presents information about configuring a router, and Chapter 10, which deals with security.

## Connecting a LAN via a Smart Screening Firewall

A simple filtering router has to decide whether to accept or reject a datagram based on the contents of that one datagram. Because the router does not

---

[1]Screening firewall products (such as Check Point Software's FireWall-1) get smarter with every release and can supervise the activities of dozens of different applications.

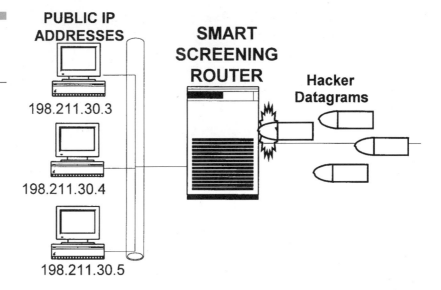

**Figure 5.2**
Connecting via a
smart screening
firewall.

remember what happened earlier, it does not have enough information to make smart judgments. A smart screening firewall remembers information about each ongoing communication.[1]

For example, in Chapter 4 we discussed the Domain Name System (DNS), which translates between computer names and addresses. A hacker might try to choke your network with junk by streaming in fake responses to non-existent DNS queries. A dumb filtering router would let these messages through because it would assume that they corresponded to earlier queries. A smart screening firewall would throw away these messages because it would know that the responses did not match any current queries. The smart firewall also would log information about this attempt to vandalize your network, or even generate an alert to management staff. Figure 5.2 shows how a site can be connected via a smart screening firewall.

# Network Address Translation Router

The supply of IP addresses is dwindling and it is becoming difficult to get a block of addresses that is convenient to use. All of the ins and outs of this situation will be discussed in Chapter 7. A Network Address Translation (NAT) router can translate between internal addresses and a small pool of

**Figure 5.3**
Connecting via a
NAT router.

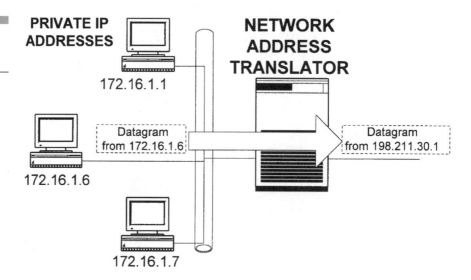

Internet addresses, or even to a single Internet address. Internet systems will not see the actual addresses of your systems. Figure 5.3 illustrates the action of a NAT Router. Network Address Translation is available in stand-alone products or as part of the capabilities of a smart screening router.

# Connecting a LAN via an Application Proxy Firewall

There is another alternative that can add protective armor to a LAN. You can connect to the Internet via an *application proxy firewall*. There are lots of application proxy firewall products that are available.

As Figure 5.4 shows, when a LAN client requests a Web page or a file on the Internet, the client opens a connection to the proxy firewall and passes its request to the proxy. The proxy opens a separate connection to the target site on the Internet, obtains the desired item, and passes it back to the client. An internal client never communicates directly with an Internet host. The proxy is the only computer that has communications sessions with Internet hosts.

Proxies have a lot of advantages:

■ A proxy provides excellent security because it prevents Internet users from accessing your LAN. The proxy is the only system that is visible to the outside world.

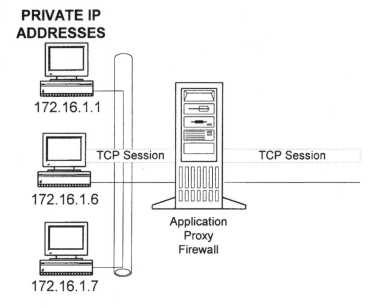

**Figure 5.4**
Hiding clients behind a proxy firewall.

- Many proxy products can screen out viruses and other potentially dangerous executables, including Java applets or *ActiveX*.

- The proxy also can be used to control which local users can access the Internet, and can limit the applications and destinations that they can select.

You might think that putting a proxy in the way will slow your Internet response time to a crawl. In fact, proxies are fairly fast and efficient. In fact, a proxy might even speed up access to popular Web pages because the proxy server will cache them onto its own disk, where they can be retrieved quickly.

There are many proxy products available with a wide range of prices, performance capabilities, and features. At the time of writing, there also are a number of free proxy packages. There is a free proxy package for Unix systems called the *Trusted Information Systems (TIS) Toolkit*. This can be obtained from Trusted Information Systems (*http://www.tis.com/*).[2] TIS also sells a firewall product called Gauntlet.

The Microsoft proxy server for NT Server provides TCP/IP and NetWare clients with access to World Wide Web, file transfer, mail, Internet news, Real Audio, VDOLive, and Internet Relay Chat application services. Finally, several free Web servers include proxying capability.

---

[2]The TIS toolkit can be retrieved directly from its file transfer site, *ftp://ftp.tis.com/pub/firewalls/toolkit/fwtkvl.3.tar.Z*.

**Figure 5.5**
NetWare desktops
and gateway server.

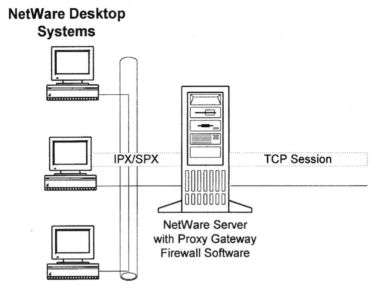

**NetWare Desktop Systems**

IPX/SPX

TCP Session

NetWare Server
with Proxy Gateway
Firewall Software

# Proxies and NetWare

There is another great advantage of proxies. There are proxy products that provide Internet access to NetWare clients, so that you do not need to install TCP/IP on the NetWare desktop computers.

The proxy in Figure 5.5 enables users to run browsers and other TCP/IP applications on NetWare PCs.

In Chapter 1, we explained that the reason you can install any Internet application you like on a Windows desktop is because software developers write their applications to run on top of the Windows Socket Programming-Application Programming Interface (Winsock-API).

The trick in allowing Internet applications on a NetWare desktop is to install a small, special Winsock library of programs on the desktops, as shown in Figure 5.6.

This Winsock convinces the Windows TCP/IP applications that they are running on top of TCP/IP. In reality, data is forwarded to a NetWare server where it is repackaged for TCP/IP transmission and forwarded to the Internet via a proxy connection.

You can get this functionality by adding gateway software to one of your NetWare servers. Alternatively, you can purchase a gateway box that is shipped with all of the required software preinstalled.

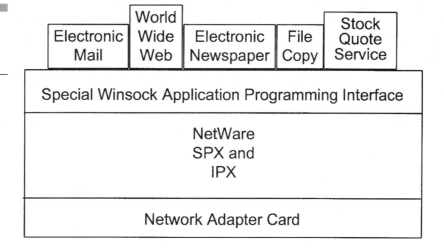

**Figure 5.6**
Running Internet
applications on top
of NetWare.

# Setting Up a Perimeter LAN

When a medium- to large-sized site connects to the Internet, an intermediate LAN, called a perimeter LAN or demilitarized zone, often is used to separate internal resources from the Internet. Figure 5.7 shows one way that a perimeter LAN could be set up. The only computers in the figure that communicate directly with Internet systems are a mail exchanger gateway that relays mail to and from the Internet and an application proxy firewall. Internal users must access the Internet via this proxy firewall.

So far, our focus has been on the equipment that connects a site to the Internet. In the sections that follow, we will survey the different link technologies that can be used between a site and an Internet Service Provider's network.

**Figure 5.7**
Using a perimeter
LAN as a buffer zone.

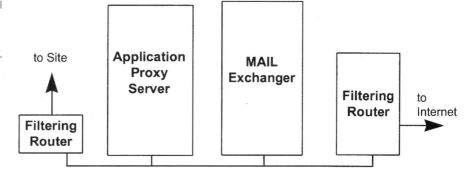

# Connecting a LAN via POTS Dial-Up

If you have a small LAN and all that you want to do is send and receive electronic mail and support occasional Web browsing, you can use an ordinary "Plain Old Telephone Service " (POTS) dial-up account for low-cost entry to the Internet. Low-cost modems that operate at up to 56 kilobits per second[3] (Kbps) are making this a feasible solution. You can obtain a low-cost dial-up router, or configure a Windows NT server or low-end Unix system to act as a router or as a proxy firewall.

Some ISPs will reserve a modem for you at their site so that you can be sure of connecting whenever you wish. You can have guaranteed access throughout the business day or pay a little more for full-time, 24-hour-per-day access. Your router or proxy firewall can be configured to remain connected throughout the day, or else can connect on demand when a user needs access to the Internet.

# Connecting a LAN via Dial-Up ISDN

If you find that POTS dial-up is too slow but you are not ready to pay several hundred dollars per month for a permanent connection, then the digital dial-up service provided by an Integrated Services Digital Network (ISDN) dial-up may be just what you need.

There is another use for ISDN. A site that has a permanent circuit can use ISDN as a backup in case the permanent circuit fails. Many routers designed for use with a permanent high-speed connection include an extra connector suitable for ISDN.

ISDN is a dial-up service that is based on digital technology. It provides high-quality digital data transmission at 64 or 128 kilobits per second. If you are interesting in using ISDN for Internet access, you need to:

■ Check whether your telephone company supports the service and can deliver it to your premises. If you are too far away (>18,000 feet) from your telephone central office you will be out of luck.

---

[3]Recall that the telecommunications world and computer world use the term "kilo" differently. A telephone transmission speed of 64 kilobits per second means 64,000 bits per second. Computer users refer to quantities of memory or disk storage in powers of 2. 64 kilobytes means $2^{16}$ or 65,536 bytes.

- Check whether your chosen ISP supports the service.
- Compute the likely cost for the service carefully.

ISDN equipment is widely available and quite inexpensive. For a modest price, you can get an ISDN card for a proxy server or routing PC. If you prefer to use a dedicated router, there are excellent, low-cost ISDN routers on the market. Many ISPs support ISDN connectivity at a price that is only slightly higher than the rate for ordinary dial-up.

However, there are some real telephone costs for ISDN users. There may be a hefty setup charge. The monthly line charge is usually higher than an analog dial-up (POTS) line fee. The biggest expense results from the fact that most telephone companies charge for ISDN calls by the minute, even if the calls are local. Fortunately, ISDN calls are set up very quickly—often within 2 to 4 seconds. This makes it practical to set up a connection on demand only when users need Internet access. The greatly improved throughput may be worth the investment.

To minimize per minute costs, you can configure your router to automatically drop the call during periods of inactivity.

# ISDN Problems

There occasionally are headaches associated with setting up ISDN. Some telephone companies are short of staff with ISDN know-how, leading to long waits to get service installed. Some telephone wires are in poor condition, either making it impossible to set up the service or making the service unreliable.

ISDN "standards" are not very standardized, and there are lots of potential incompatibilities. Your equipment will need to communicate with an ISDN-capable telephone switch. Unfortunately, telephone switch vendors implemented ISDN digital dial-up in many different ways. Not all user equipment works with all switches. At the very least, the configuration of your equipment will be different depending on the telephone switch in use at your central office.

If you are not familiar with ISDN, then it is a good idea to choose your Internet Service Provider before you go out and buy an ISDN card or a router. The ISP will have worked with one or more telephone companies and will have a list of recommended equipment. Sticking to a recommended product will probably save you a ton of headaches. Often, an ISP will be happy to provide and install a suitable ISDN router as part of its start-up service, and the cost for doing this is quite reasonable.

# ISDN Terminology

There is a lot of special terminology associated with ISDN , and understanding it will help you deal with your telephone company and Internet Service Provider.

The telephone system originally was based on *analog* transmission. This means that sounds were represented by variable levels of current flowing through copper wires. For many years, the telephone system has been a mixture of analog and digital technologies. As shown in Figure 5.8, information is transmitted in analog form between your telephone and your telephone central office (also called your "local exchange"). If you are calling a destination that is attached to a remote central office, the call is transformed into digital information. At the destination central office, the call is transformed back into analog form and delivered to the destination. The copper wires between your phone and your central office are called the *local loop*.

This does not look like a sensible way to run a network in today's digital world. Since 1984, telephony companies around the world have been dreaming and scheming to change the telephone system into an end-to-end digital network. They have invested a huge amount of time developing the Integrated Services Digital Network (ISDN) specifications, which are aimed at doing this.

The basic idea is simple. By using special equipment at your site and at the central office site, you will be able to transform your local loop into two 64-kilobit-per-second call channels (called bearer or "B" channels) plus a

**Figure 5.8**
Analog local loop and digital backbone.

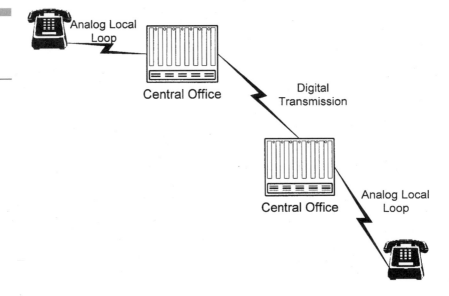

**Figure 5.9**
Bearer and signaling channels.

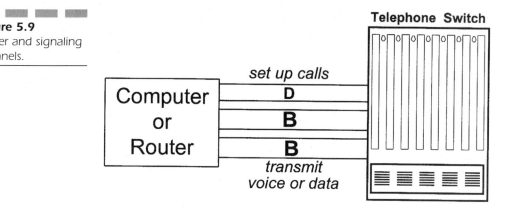

16-kilobit-per-second signaling channel (called the "D" channel) that is used to set up and terminate calls. The configuration is shown in Figure 5.9. Each 64 Kbps channel can be used for either voice or data.

Many routers support the ability to combine the two 64-Kbps channels into a single 128-Kbps channel. Of course, when you do this, you will have to pay the telephone company for two calls because you are using both channels.

Because of the way that some older U.S. telephone equipment works, your channels may actually be restricted to 56 Kbps. Nonetheless, ISDN channels will give you a lot more throughput than you can get with an ordinary POTS dial-up modem. The digital technology that is used is more reliable, has far lower error rates, and can operate at full speed in both directions.

Your telephone provider may use the term *NT1* when discussing your ISDN installation. This is a network termination box that the telephone company will screw into the wall. At this box, your four-wire telephone interface will be switched over to two-wire twisted pairs used by the telephone company.

Chapter 9 provides some information that you will need in order to configure a router to use ISDN. To get lots of information about ISDN technology and products, visit Dan Kegel's ISDN page at:

*http://alumni.caltech.edu/~dank/isdn/*

If you want to learn a lot more, then see Kessler and Southwick's *ISDN: Concepts, Facilities, and Services* (McGraw-Hill).

# ADSL

ADSL (Asymmetric Digital Subscriber Line) is called asymmetric because a large amount of information can be sent in one direction and a smaller amount can be sent the opposite way. For example, ADSL might be used for Internet access from a user's browser. A lot of information would come down the telephone line, and a smaller amount would be sent up the line. This matches the needs of a typical Web browser session quite well.

ADSL technology actually provides three concurrent channels. One can be used for an ordinary telephone call. A second provides one-way transfer of a lot of data. The data rate will vary according to length, setup, and quality of the local loop. Rates of 6 megabits per second are possible. The third channel supports two-way data transfer at rates of up to 640 kilobits per second.

The ADSL modem builders are still at work on their products, and some have achieved higher data rates. On the other hand, when installed on less than optimal wire, you might not be able to get anywhere near this level of throughput.

At the time of writing, ADSL was being used in trials and products were at the initial roll-out stage. Figure 5.10 illustrates the ADSL architecture. Modems at the end of the twisted pair loop cooperate to create the three channels: a one-way high-speed data channel to the subscriber, a lower-speed full-duplex channel, and a third channel for ordinary telephone service. You can make an ordinary telephone call at the same time that ADSL data service is in use.

Note that some very high-speed facilities will need to be in place between your central office and your ISP's network. ISPs that are telephone companies or else have partnerships with telephone companies are well-positioned to offer special services that are delivered to the customer via this technology.

**Figure 5.10**
ADSL three-channel architecture.

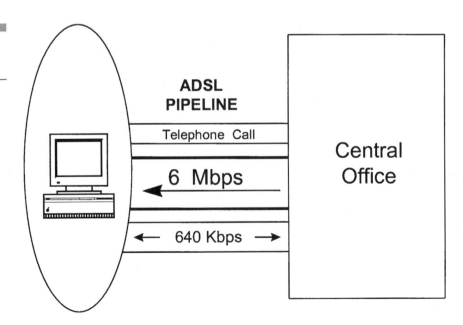

Dan Kegel has an ADSL page at:

*http://alumni.caltech.edu/ ~dank/isdn/adsl.html*

# Other DSLs

Several other digital subscriber line technologies have been developed, including:

- HDSL (High bit rate Digital Subscriber Line), which delivers T1 and E1 across copper local loops.
- VDSL (Very high bit rate Digital Subscriber Line), which under ideal circumstances might deliver 52 Mbps downstream and 2 Mbps upstream over a single twisted pair.

The DSL technologies are being rolled out at the time of writing, and are very likely to become popular choices.

# Leased Lines

If you want reliable throughput with a contract that guarantees service levels, then you may want to start pricing out a digital leased-line connection. ISPs offer leased-line connections at speeds ranging from 56 kilobits per second to 45 megabits per second.

Most often, a router is used for a leased-line connection. However, a serial communications card can be installed into a firewall system that interfaces directly to a leased line. In the discussion that follows, we will assume that the LAN (or larger site) is connected via a router.

# CSU/DSU

When you dial the Internet you use a modem. The modem converts digital data transmitted by your computer into analog "tones"[4] that are transmitted across your local loop.

---

[4]You probably have heard your modem sending and receiving some of these tones during call setup.

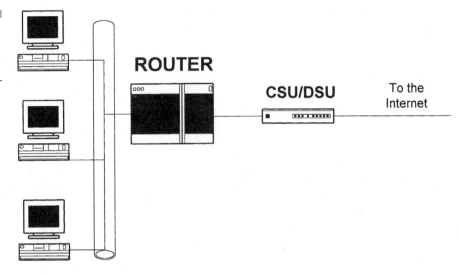

**Figure 5.11**
Connecting to a
digital telephone line
via a CSU/DSU.

In contrast, when you sign up for a digital leased-line service, data is transmitted in digital form from end to end. Instead of using a modem, you interface to the digital telephone system by means of a box called a Channel Service Unit/Data Service Unit (CSU/DSU).

High-speed digital telephone transmission depends on clock synchronization that is used to determine when bits begin and end. The DSU synchronizes the clocking used to transmit bits with telephone network clocking. The CSU provides the physical interface to the telephone network and performs low-level functions like setting transmit and receive levels. In Figure 5.11, a router interfaces to a telephone line via a CSU/DSU. The CSU/DSU is shown as an external box, but for some routers, it can be integrated into the router's chassis.

There are two main types of CSU/DSU. One is used for a 56- or 64-kilobit-per-second line. Another is used for access at rates ranging from 128 kilobits per second to more than 1.5 megabits per second. A special DSU is used for high-speed access at rates ranging from 3 megabits per second to 45 megabits per second.

# ISP Installation

Many ISPs will provide you with a router and CSU/DSU, and will order and supervise the installation of the telephone line. There are good reasons to accept the ISPs choice of router and pay for their installation service. The router that is chosen will be one that the ISP staff knows well and they will be able to set it up quickly and correctly. ISPs choose routers that support the Simple Network Management Protocol (SNMP) so that they can troubleshoot router and line problems.

## 56-Kilobit Leased Lines

56- or 64-kilobit-per-second digital leased lines are the venerable workhorses of data communications and are available just about anywhere. Once the most prevalent form of data communications, 56- and 64-Kbps lines are seriously challenged by lower-cost frame relay services, which will be discussed later.

## Fractional T1, T1, and Dual T1

T1 technology carries a bundle of twenty four digitized telephone calls. Each call occupies a 64-kilobit-per-second channel. T1 has become the standard technology for leased-line data communications, providing over 1.5 megabits per second.

ISPs allow you to contract for convenient fractions of this bandwidth: e.g., 128, 256, 385, or 512 kilobits per second. However, the initial price of entry might be high, depending on your location. Before you can use fractional T1, you have to bring a full T1 line to your premises. This might cost you several thousand dollars. Once this has been done, you will pay for whatever fraction you need. A positive feature is that you can increase your bandwidth as needed without changing the equipment that you use.

Some providers offer a very flexible billing scheme. They open up a T1 pipeline to your site and let you use as much bandwidth as you need, but then apply a usage averaging formula when they bill you. It is a good idea to examine the formula carefully before committing to this service.

It is fortunate that cheap T1 access is becoming more and more prevalent. T1 access is highly recommended if providing your users with Internet access will bring a solid benefit to your organization, or if you are going to host a serious Internet Web server from your own site. Some businesses are upgrading to dual T1 offerings that combine two T1 channels into a 3-megabit-per-second pipeline.

# E1

Outside of North America and Japan, phone systems use E1 carriers instead of T1. E1 carriers combine thirty-two 64-Kbps channels. Two channels are reserved for signaling and alignment functions. Hence, there are $30 \times 64,000 = 1,920,000$ bits of payload. Many vendors build equipment that can be used in either T1 or E1 environments.

# Frame Relay

Frame relay is a technology that provides cost benefits and flexible band width options. A frame relay service provider builds a wide area network that can connect its subscribers to many locations. A single line connects an end-user site to the frame relay network. Multiple circuits can fan out to a variety of other locations that also are connected to the frame relay net. A frame relay circuit costs a lot less than a leased line because high-bandwidth links within the frame relay network are shared by many customers.

Today, many ISPs connect to a frame relay network and support frame relay connections to their customers. Large ISPs even operate their own frame relay networks. Often, the ISP will provide and set up all of the equipment needed, get the service going, and monitor and manage your connection to the frame relay network. You could combine access to the Internet with private circuit access to other business sites, as shown in Figure 5.12.

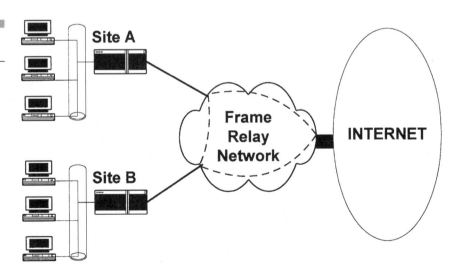

**Figure 5.12**
Frame relay circuits.

Depending on the throughput that you need, either a 64-Kbps[5] or a T1 (1.54 megabit per second) line connects you to the frame relay network. Once that connection is in place, you will pay for whatever level of bandwidth you desire. An advantage of using a T1 access line is that you can increase the service level according to your needs without any change in equipment. (Alternatively, you could use some of the T1 bandwidth for ordinary telephone calls.)

Frame relay offers another important advantage. Data traffic tends to be bursty. You may have contracted for 128 Kbps, but sometimes your needs will rise above that level while at other times you need less. Since you must have a T1 pipeline in order to support 128 Kbps, you physically are able to send and receive at higher rates. Your Service Provider will allow you to briefly exceed your limit as long as the average stays within contractual bounds.

# Frame Relay Disadvantages

The reason that frame relay is priced so attractively is that a frame relay network contains a mesh of switches and links that are shared by a large number of customers. If some link in the frame relay network gets very busy and becomes congested, some of your traffic might be thrown away and have to be retransmitted later. Therefore, it is important to ask your frame relay provider about the current level of network utilization. Then find out whether the provider will back up assurances of low congestion with a service level guarantee.

Another disadvantage of frame relay is that it takes some extra time for frames to be routed through the network switches. It is a good idea to obtain some response time estimates for the frame relay network. The types of switches that are used within the frame relay network make a difference; cut-through switches can move frames more quickly.[6]

# Fractional T3 and T3

Now we are entering the realm of the heavy hitters. These are powerful services for big-time Internet users who probably are hosting their own popular

---

[5]In Chapter 9, we shall see that some of these lines actually deliver only 56 Kbps.

[6]A cut-through switch can start transmitting a frame across its next link immediately, without waiting until the entire frame has been received by the switch.

Web site. T3 lines were created as a way to bundle 28 T1s together. The raw T3 transmission rate is almost 45 megabits per second.

After a hefty setup charge to bring T3 service to your premises, most providers who support this service will allow you to use a convenient fraction of the bandwidth, ranging from 3 Mbps to 45 Mbps at 3-Mbps intervals (3, 6, 9, 12, etc.)

# SMDS

Switched Multi-Megabit Data Service (SMDS) is a data transmission service originally offered by Regional Bell Operating Companies. Speeds ranging from 1.177 Mbps to 34 Mbps are available. The startup charge and monthly rates are lower than T3 charges. Another advantage is that a customer who has several sites can set up a virtual LAN that operates across a wide area. A disadvantage of SMDS is that its availability is very limited.

Figure 5.13 shows locations that connect to one another and to an ISP via an SMDS network. The figure looks quite similar to Figure 5.12, which illustrated frame relay connectivity. However, the technologies are quite different, and SMDS services usually offer higher bandwidth options than frame relay services.

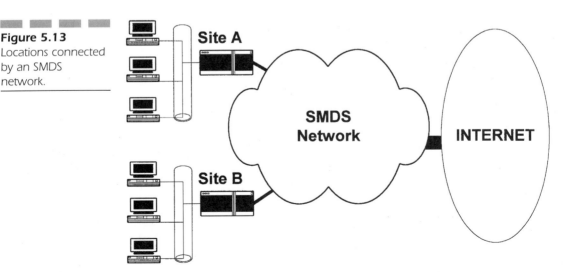

**Figure 5.13**
Locations connected by an SMDS network.

**Figure 5.14**
Connecting at 10
Mbps via bridges and
ATM.

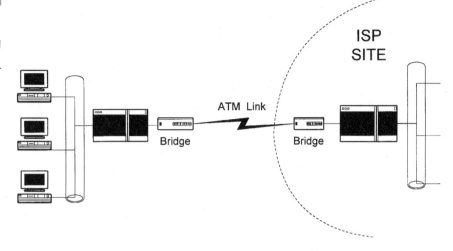

## 10-Megabit Service

Some providers offer connectivity at 10 megabits per second. Figure 5.14 shows a configuration that sometimes is used. In a typical implementation, routers at the customer and ISP sites interface to an Ethernet and to high-speed bridges. The bridges are connected by an *Asynchronous Transfer Mode* (ATM) telephone link. ATM is a technology that breaks up information into small data cells and switches these at very high speeds.

Telephone technology is expected to migrate to ATM in the long run. ATM telephone service offers many different bandwidth options including very high bandwidths. It also can support various classes of service, enabling ATM to be used for data, voice, and video. Currently, ATM service is obtainable in relatively few areas, but availability is expected to expand steadily.

This configuration in Figure 5.14 is not the only one that is used. Sometimes 10 megabits per second is supported via a connection to a Fiber Distributed Data Interface (FDDI) LAN.

## Cable Service

Some cable vendors are gearing up to provide Internet access to businesses and end users via a special cable box that includes an Ethernet interface. Subscribers would use a router or personal computer with an Ethernet interface and an Ethernet cable to connect to the cable box.

# 6

# Evaluating and Choosing a Service Provider

There are a lot of Internet Service Providers in the marketplace and Internet connectivity may look like a standard commodity, but it isn't. There are many considerations besides price that need to be factored into your provider selection. These include:

- Communications Options
- Performance
- Installation and Monitoring
- Quality of Service
- Support
- Security Services
- Application Services
- Plans for New Technologies
- One-Stop Shopping

How important these are depends on how you plan to use the Internet over the next few years. If your only use of the Internet is electronic mail transport, then you probably would not even notice an occasional service outage and your security needs can be taken care of very simply. If you are selling goods or services on the Internet, then an outage means lost revenue and possible lost customers.

# Communications Options

We examined several communications methods in Chapter 5, and you may already have selected the one that best suits your needs today. Keep in mind that transmission technologies and tariffs are changing at a dizzying rate, and look for a provider that has a good record for keeping up with new offerings. Otherwise, you may be faced with a choice between changing providers or staying with a technology that has been eclipsed or offers less than optimal value per dollar.

Changing providers is a costly and disruptive process. You have to pay a new start-up fee. Your registration information needs to be updated. You probably will have to change the IP addresses of a number of your computers. Entries in your Domain Name System directory have to be updated and one or more of your directory servers probably will need to be moved. During the changeover, your access to the Internet and even your electronic mail service may be disrupted.

# Performance

Your Internet performance will be affected by many factors. You may have a T1 line but once your traffic reaches your provider, the provider may be dumping traffic for half a dozen T1s onto a single T1 link that connects the provider to the rest of the Internet. Or your traffic may have to traverse several routers before reaching a switch that can take it onto the destination provider's network.

## Provider's Internal Network Structure

The provider's own internal network needs to be robust enough to satisfy its customers' current transit needs. It also needs to be in a continuing state of expansion. User appetite for Internet bandwidth has grown steadily over the years, and there is no sign that it is abating.

Check that the provider uses up-to-date routers that can be upgraded for higher capacity and new communications technologies. The provider should have plans for its own growth in place. Ask about their capacity planning methodology.

## Connections to Other Service Providers

Traffic will flow between your network and points all over the world. The bandwidth of the inter-provider links needs to be large enough to handle ever-increasing customer requirements.

A provider that is connected to several of the major switching points can shunt traffic directly onto many other provider networks, and will give you faster response time than a provider whose traffic has to pass through many routers and through other ISPs in order to reach switching points. (See Figure 6.1.)

## Installation and Monitoring

If keeping your Internet link up and operating efficiently is important to your organization, then you should consider using an ISP that will install and

**Figure 6.1**
Chain of routers leading to a major switching point.

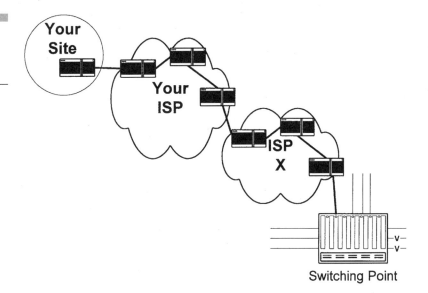

Your Site

Your ISP

ISP X

Switching Point

configure your router, take responsibility for the condition of your access line, and monitor the router and access line on a 24-hour-per-day basis. An ISP that uses management tools based on the Simple Network Management Protocol (SNMP) can watch over your site and troubleshoot router problems effectively.

This type of support service can expedite problem solving and save you from finger pointing between your own staff, the telephone company, and the Internet Service Provider. However, it is a good idea to check up on the size of the ISP's network operations staff before contracting for full service.

## Quality of Service

If high availability is important to you, find out whether the ISP has redundant routes, dual routers, and backup power supplies.

An ISP should be able to give you information on its availability record and document it with raw data on outages. You can check up on their record by talking to some existing customers. Some ISPs back up their claims with service level guarantees that offer financial reimbursements when the percentage of up time falls below a threshold (such as 99%). But be sure to get all service level guarantees in writing.

Several ISPs give you a real-time window into what is happening on their networks. They will allow you to connect to a management system at their Network Operations Center. You will be able to see for yourself whether equipment is functioning and how throughput is being maintained.

## Support

The Internet runs 24 hours per day, 7 days per week. If you are operating Internet Web or file transfer servers, or if your staff needs access outside of regular business hours, your ISP should provide support around the clock. Before you sign up, try calling the support telephone number at midnight on a Saturday night!

But for support, just being there is not enough. Your ISP should have a well-defined procedure for you to use to report and track a problem. An online trouble ticket system is best. You should be able to get information on who is handling your problem and when the work is expected to be completed.

# Security Services

Some ISPs offer no security service at all. Others can install and configure a security firewall system at your site and monitor your Internet traffic 24 hours per day. ISPs who offer good security services have personnel who are immersed in security technologies. They will be aware of new threats on the Internet and will be able to spot break-in attempts around the clock. They will audit your exposed systems, look for suspicious events, and clean up damage.

Security services can be costly so you need to assess your risks and exposures. A small PC LAN can be protected by a simple application proxy server. A banking organization needs a full blown protective service as well as additional tiger-team consultants who may find loopholes that even a good security staff may have overlooked.

# Application Services

It is routine for ISPs to operate Domain Name System directory servers, forward electronic mail to and from your site, and operate a network news server.

Most ISPs also provide Web (and optionally file transfer) hosting service. Placing your Web information on their network can be advantageous. Some ISPs locate their "farm" of Web servers on a network with a high-bandwidth attachment to a switching point that supports quick access from anywhere. Because of economies of scale, an ISP can use superior equipment, run redundant servers in case of a failure, and provide a 24-hour-per-day staff to oversee and protect all of the systems.

Outsourcing your Web site also allows the link to your site to be dedicated to giving Internet access to your users. It simplifies the job of securing your site, since you can block external users from connecting to your systems.

There can be some disadvantages to outsourcing your Web site. It can be costly. You need to make sure that you will not lose the ability to update your information whenever you wish. And, if you are performing business transactions rather than just serving out Web pages, internal expertise may be needed in order to run your business applications.

# Plans for New Internet Technologies

There are emerging technologies that will be the basis of new Internet services. However, you will not be able to use these services unless your ISP supports the enabling technologies. We'll present two emerging technologies in the sections that follow.

## Multicasting

There already is a demonstrated demand for up-to-the-minute news or stock quotations delivered to the desktop. This type of information service is best supported via IP multicasting, which can deliver data to thousands of desktops efficiently. Multicasting also is an appropriate technology for online conferences.

Multicasting standards exist and the technology has been tested and proven. When a computer joins ("tunes in") to a multicast, routers cooperate to make sure that the computer receives all of the datagrams for that multicast. Figure 6.2 illustrates how a single multicast message is replicated along paths to subscriber computers, but is not delivered to non-subscribers.

**Figure 6.2**
Delivering a multicast message.

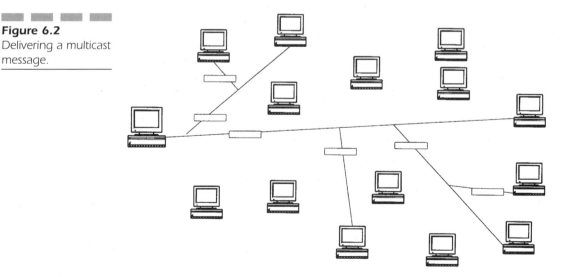

Before its users can participate in multicasts, an ISP will need to install and maintain multicast software in its routers. This also will require staff that understand the technology and can dedicate time to cooperating with peers at other ISPs.

## RSVP

Reservation Setup Protocol (RSVP) is a router feature designed to provide a guaranteed quality of service for particular sessions. For example, RSVP will be used to support real-time applications such as video.

The way that RSVP works is that the computer that will be the data receiver initiates a request for reserved bandwidth. The receiver sends its request to a neighbor RSVP router. That router checks its current status and tells the receiver whether it has enough unused bandwidth resources to satisfy the request. If the answer is yes, the router forwards the request to the next router on the path towards the sender. If all reservations succeed, then a reserved pipeline has been set up between the sender and the receiver.

# One-Stop Shopping

Internet connectivity is becoming just one item in a shopping basket that includes telephone access, internal leased lines, and private connections to suppliers and customers. There often are cost and convenience benefits in using a single provider for several services.

# Price and Service Negotiations

If you are near an urban area, many ISPs will be vying for your business. It is a good idea to take the time to meet with several of them and analyze their offerings. You can negotiate favorable rates on a multi-year contract. If you need high availability, you will be in a good position to request service level guarantees backed up by rebates. But be sure that all agreements are in writing!

**Figure 6.3**
A sample service
agreement.

## Service Agreement
### Internet Access via Frame Relay

The undersigned Subscriber requests VNET Internet Access ("Company") to provide Internet Access and Frame Relay Service at the following subscriber's location(s):

_____

_____

_____

Important provisions relating to Internet Access and Frame Relay Service are set forth herein:

The Company will furnish, install, maintain and provide maintenance for Internet Access and Frame Relay Service ("Service") in accordance with lawfully filed tariffs. The tariffs provide the basis for this Agreement with the Subscriber. The Agreement period shall begin the day Frame Relay Service is installed.

The Subscriber agrees to pay the Company for the provision of the Internet Access and Frame Relay Service. The Service shall be offered for variable rate periods of 12 months to 60 months. This monthly rate will continue for the elected service period and will not be subject to Company initiated change during such period.

The monthly rates for Internet Access and Frame Relay Service in effect at the time the Service is installed and/or as of the service order application date will be in effect until the expiration of the service period chosen by the Subscriber. Other rates applicable to other services provided by the Company, including but not limited to, feature charges and private line channel services that are connected to Frame Relay Service, may be increased during this period.

The service period for this Agreement shall be _____ months. The rates and charges, per month, for items under this Agreement are:

| Service | Non Recurring Charges | Recurring Charges |
|---|---|---|
| Internet Access & Frame Relay | | |
| | | |
| Totals: | | |

In the event that any item of the Service is terminated prior to the expiration of the service period, the Subscriber shall pay a Termination Liability Charge. The Termination Liability Charge is determined by multiplying the number of months remaining in the contract payment period by the contracted monthly rate for Service by 60 percent.

At the expiration of the service period, the Subscriber may continue the Service according to renewal options to be negotiated at that time. If the Subscriber does not elect an additional service period or does not request discontinuance of service, then the above Service will be continued at the monthly rate then currently in effect for month-to-month rates. Service periods may also be renewed prior to expiration in accordance with regulations and rates then in effect.

Suspension of service is not permitted for Internet Access or Frame Relay Service.

The Subscriber agrees to pay any added costs incurred by the Company due to a Subscriber initiated change in the location of the Frame Relay Service prior to the time it is placed in service.

In the event the Service requested by the Subscriber is canceled prior to the establishment of Service, but after the date or ordering reflected herein, the Subscriber is required to reimburse the Company for all expenses incurred in handling the request before the notice of cancellation is received. Such charge however is not to exceed the sum of all charges which would apply if the work involved in complying with the request had been completed.

**Figure 6.3
continued**

A sample service
agreement.

The Subscriber may arrange to have existing Service under this Agreement moved within the same premises. Subscriber agrees to pay a non-recurring charge based upon the estimated cost of such rearrangement without interruption or change in the monthly rates.

Service may be transferred to another Subscriber at the same location upon prior written concurrence of the Company. The new Subscriber to whom the Service is transferred will be subject to all tariff provisions and equipment configurations currently in effect for the present Subscriber.

This Agreement is effective when executed by the Subscriber and accepted by the Company and is subject to and controlled by the provisions of this Agreement, including any changes therein as may be made from time to time.

Accepted By:_____

Title: _____ On the _____ Day of _____ ,1997

**VNET Internet Access**

Accepted By: _____

Title:_____ On the _____ Day of _____ , 1997

A basic Internet Service Provider contract is usually not a complicated document. Figure 6.3 shows a sample contract from VNET Internet Access (*www.vnet.net*), which offers connectivity services in many U.S. cities and connections to several other countries. VNET's home office is based in Charlotte, North Carolina.

# 7

# IP Addresses

Most of us are familiar with Internet names such as *www.whitehouse.gov* or *home.netscape.com*. These names now appear in our newspapers and magazines, are mentioned on TV, or even are displayed on billboards and on the sides of buses.

But just as you cannot place a telephone call to someone without knowing their telephone number, your computer cannot communicate with a destination host without knowing its IP address. Unfortunately, IP addresses are not as well designed as telephone numbers are.

Telephone numbers have country codes and area codes. These give a *hierarchical structure* to the phone system. By this we mean that the structure of our telephone numbers simplifies the job of routing a call. A telephone switch can tell that you are calling a foreign country by examining the first 3 digits that you dial. (In the United States, we dial 011 for international calls.) The next digits enable the switch to route an international call to the correct country. The job of examining the remaining digits to identify the destination area can be delegated to a switch in the target country.

This type of structure was not built into IP addresses. When IP addresses originally were defined, their designers believed that addresses would never be needed for more than a few dozen IP networks. No one knew that IP addresses eventually would be used to identify millions of computers spread across thousands of networks located all over the world.

Assigning blocks of IP addresses to organizations has become a big job. It used to be done by one central address registry. Now the work has been farmed out to regional registries and Internet Service Providers. In this chapter, we are going to describe how this works. We'll also present the classic IP numbering scheme, and see how address allocation procedures have been changed so that current address assignments follow a hierarchical address structure that improves the efficiency of Internet routing.

The supply of unique IP addresses is dwindling. This chapter will explain how you can use private enterprise addresses, proxy servers, and network address translation servers to survive the address crunch. You also can conserve IP addresses by using the gateway technique described in Chapter 5 to connect your NetWare desktops to the Internet.

Finally, this chapter includes information about subnets, subnet masks, and other topics needed to round out an understanding of IP addresses.

# How IP Addresses Are Used

If you could take an X-ray of the Internet, you would see small chunks of data—datagrams—racing from sources to destinations. Each datagram has a header that identifies where it is going and where it came from—namely, its destination and source IP addresses. The destination IP address is used to route a datagram to its destination.

# Writing IP Addresses in Binary and Decimal

We already have seen many examples of IP addresses in this book. For example, in Chapter 1, we discovered that the computer named *www.white-house.gov* has IP address 198.137.240.91.

Although we are used to seeing IP addresses written as a series of decimal numbers separated by dots, the underlying addresses actually are 32-bit binary quantities. A 32-bit address is translated to its dotted decimal format by breaking the 32 bits into four 8-bit bytes and then writing each of these bytes as a decimal number. Table 7.1 shows some of the smallest and largest numbers that can be expressed using 8 bits.

**TABLE 7.1**

Representing Bytes
in Binary and
Decimal

| Binary Byte | Decimal |
|:---:|:---:|
| 00000000 | 0 |
| 00000001 | 1 |
| 00000010 | 2 |
| 00000011 | 3 |
| 00000100 | 4 |
| . . . | |
| 11111101 | 253 |
| 11111110 | 254 |
| 11111111 | 255 |

We can see that 255 is the biggest decimal number that will appear in an IP address, and therefore IP addresses range from 0.0.0.0 to 255.255.255.255.

Appendix A of this book explains how binary numbers are converted to decimal numbers. Appendix A also contains a complete conversion table that converts binary bytes to decimals ranging from 0 to 255.

# IP Address Format

Every IP address starts with a *network number*. An address registrar assigns addresses to an organization by giving it one or more network numbers.

As Figure 7.1 illustrates, there are different sizes used for the network number part of an IP address. The remainder of the address is called the local part.

# Address Classes

Traditionally, there were three classes of IP addresses:

■ Class A addresses were used for very large networks. The first byte was the network number part of the address. The organization could use the remaining 3 bytes any way that it wished. The address space contained 16,777,216 numbers.

**Figure 7.1**
Network numbers
and local parts.

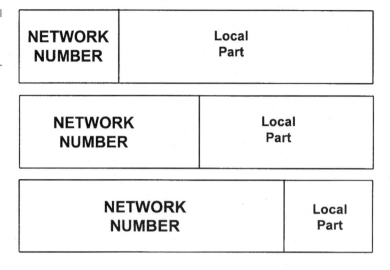

- Class B addresses were used for medium-sized networks. The first two bytes made up the network part of the address. The organization could use the remaining 2 bytes any way that it wished. The address space contained 65,536 numbers.
- Class C addresses were (and still are) used for small networks. The first three bytes made up the network part of the address. The organization can use the remaining byte any way that it wishes. The address space contains 256 numbers.

# Recognizing Classes

You can identify the class of an address by looking at the first number in dotted decimal form of the address. Table 7.2 summarizes the address class characteristics.

## Class A

If the first of the four address numbers is in the range 0 to 127, then it introduces a Class A address. The first number is the network part of the address. For example, Digital Equipment Corporation has Class A network number 16. System *ns1.pa.dec.com* has address 16.1.240.41.

**TABLE 7.2**

Address Class
Characteristics

| Class | First number | Length of Network Part | Number of Addresses |
|-------|-------------|------------------------|---------------------|
| A | 0–127 | 1 | 16,777,216 |
| B | 128–191 | 2 | 65,536 |
| C | 192–223 | 3 | 256 |

Digital Equipment Corporation is free to assign the three numbers on the right to hosts in any way that they wish.

## Class B

If the first number is in the range 128 to 191, then it introduces a Class B address. In this case, the first two numbers make up the network part of the address. For example, Yale University uses the Class B network number 130.132 (along with several other network numbers). System *mail-relay1.cis.yale.edu* has Class B address 130.132.21.199. The network part of this address is 130.132. Yale can assign the two numbers to the right in any way that they wish.

## Class C

If the first number is in the range 192 to 223, then it is a Class C address. The first three numbers make up the network part. For example, Data-Tech Institute uses the Class C network number 204.32.6. System *www.datatech.com* has IP address 204.32.6.220. The network part of this address is 204.32.6. Data-Tech controls the last number in the address.

## Class Inefficiencies

As you can see, there is a big difference in the sizes of Class A, B, and C networks. This led to very inefficient assignments of addresses. For example, in the early years of the Internet, organizations with a couple of thousand hosts could get Class B addresses. These wasteful allocations could not continue. Addresses are now assigned in a "classless" manner that fits the size of an address block to the real needs of the customer. We'll explain how this works a little later. First, let's talk about how addresses are obtained today.

# Obtaining Blocks of Addresses

Initially, all blocks of addresses were given out by a single, central registration service in the United States. Using a central registrar and handing out Class A, B, or C network numbers worked well when the Internet was a small community of academic and research sites. It was badly flawed for global operation:

- Eventually, there was too much work for a single registry to handle.
- Defining only three address block sizes led to very wasteful use of addresses.
- IP addresses lacked a hierarchical structure similar to the one that the telephone system has (country codes and area codes). Some type of hierarchy was needed in order to simplify the job of routing data.

The old, inefficient method could not support today's Internet.

# Registration Centers

As the Internet exploded in size, it was clear that the job of assigning addresses had to be farmed out to multiple administrations and handled in a hierarchical fashion. The first step in reforming address registration was to turn top-level responsibility over to the Internet Assigned Numbers Authority (IANA). The IANA still has this oversight responsibility. Their World Wide Web site is *http://www.isi.edu/iana/*.

The hierarchy used today is fairly loose. There are three major regional address registries. These are the same registries that were described in Chapter 4. There is a plan to create a new non-profit organization (the American Registry for Internet Numbers) that will carry out the address allocation functions currently performed by the Network Solutions InterNIC.

Currently, the Registration InterNIC oversees addresses for North, Central, and South America. The InterNIC also handles special cases, directly issuing addresses to organizations that have extremely large networks. Reseaux IP Europeens (RIPE) is the European address registration center. The

Asian-Pacific Network Information Center (APNIC), located in Tokyo, oversees addressing for Asia and the Pacific Rim countries. The IANA and the InterNIC have assigned large blocks of addresses to these registries.[1]

Internet Service Providers obtain their addresses from an appropriate regional registry. Sometimes an ISP will delegate part of its address space to a smaller ISP.

# End-User Addresses

Internet Service Providers have primary responsibility for supplying end-user organizations with addresses. Except for very large enterprises or organizations that connect to more than one ISP, end-user sites must obtain their addresses from an ISP.

Let's take a look at an example. The U.S. SprintLink Service Provider has been assigned many blocks of addresses. We can view their address allocations by doing a *Whois* lookup on SprintLink.[2] The list is very long, so we will just display excerpts:

```
> whois -h rs.internic.net Sprintlink
Sprint (NET-SPRINT-INNET5)   SPRINTLINK                    144.228.0.0
. . .
SprintLink (NETBLK-SPRINT-N) SPRINT-N 206.158.0.0 - 206.158.255.255
SprintLink (NETBLK-SPRINT-W) SPRINT-W 206.159.0.0 - 206.159.255.255
...
```

Sprint can assign addresses to their customers from their blocks of numbers. Below, we look up 206.159.0.0, which is a number within the Sprint address space. We find that Sprint has assigned the block of network numbers 209.159.0-209.159.63 to a smaller ISP called Wolfe Internet Access.

```
> whois -h rs.internic.net 206.159.0.0
SprintLink (NETBLK-SPRINT-W) SPRINT-W 206.159.0.0 - 206.159.255.255

Wolfe Internet Access, L.L.C. (NETBLK-SPRINT-CE9F3F) SPRINT-CE9F3F
                              206.159.0.0 - 206.159.63.0
```

When we look up 206.158.2.0, we find that this network number has been assigned to a company called Automated Data Systems.

---

[1]You can perform a *Whois* lookup on RIPE or APNIC to view numbers that they have been allocated.

[2]As noted in Chapter 4, if we want to make queries about networks in Europe or the Pacific Rim, we would need to send queries to *whois.ripe.net* or *whois.apnic.net*.

```
> whois -h rs.internic.net 206.158.2.0
Automated Data Systems, Inc (NET-SPRINT-CE9E02)
   618 Franklin Square
   Michigan City, IN 46360
   US
   Netname: SPRINT-CE9E02
   Netnumber: 206.158.2.0
```

# Classless Addresses

The old address allocation scheme assigned addresses in a very wasteful manner. Today, an organization is given a block of addresses whose size closely matches its needs (allowing for growth over the next year or two). Currently, most blocks are taken from the Class C address space. We still think of the blocks as being made up of a batch of consecutive Class C addresses. This is a convenient way to think of the addresses when we write them down using the dotted decimal notation.

Table 7.3 shows typical address assignments. In the previous section, we saw that Wolfe Internet Access has control of the equivalent of a block of

**TABLE 7.3**

*Address Allocations*

| Require at Most | Number of Class C Networks | Number of Bits in Prefix |
|---|---|---|
| 30 addresses | 1/8 | 27 |
| 62 addresses | 1/4 | 26 |
| 126 addresses | 1/2 | 25 |
| 254 addresses | 1 | 24 |
| 510 addresses | 2 | 23 |
| 1022 addresses | 4 | 22 |
| 2046 addresses | 8 | 21 |
| 4094 addresses | 16 | 20 |
| 8190 addresses | 32 | 19 |
| 16382 addresses | 64 | 18 |
| 32766 addresses | 128 | 17 |
| 65534 addresses | 256 | 16 |

64 network numbers ranging from 206.59.0 to 206.159.63. As we can see in the table, this block contains over 16,000 addresses.

The table also shows allocations that are a fraction of a Class C network. ISPs have been told to conserve addresses. They will assign a block of addresses that is a fraction of a Class C block to a very small site.

These new blocks can be viewed as addresses whose network numbers are a specified number of *bits* instead of bytes. The number of bits in each network number prefix is shown in Table 7.3.

# Bits versus Class A, B, C

The material in this section gives you a different way of looking at network addresses and may help you to communicate with the technical staff at your ISP.

An Internet address consists of 32 bits. Computers are very happy working with bits. People generally are not. Addresses were broken into 4 bytes and then translated to 4 decimal numbers to make people happy. The original Class A, B, and C addresses were designed using byte boundaries in order to keep people happy.

But now that the Internet address crunch has arrived, arbitrary bit boundaries are used. The blocks of contiguous Class C addresses that we talked about in the previous section correspond to network numbers that occupy some number of bits. Let's look at some examples.

Automated Data Systems has been assigned the traditional Class C network number 206.158.2.0. The network number consists of the 24-bit prefix:

11001110    10011110    00000010

Wolfe Internet Access has been assigned the block 206.159.0.0–206.159.63.0. Every one of these numbers starts with the 18-bit pattern:

11001110    10011111    00

This 18-bit prefix is viewed as Wolfe's network number.

SprintLink has been assigned many blocks of addresses, including 206.158.0.0–206.159.255.255. Every one of the numbers in this block starts with the 15-bit pattern:

11001110    1001111

This 15-bit prefix is viewed as the network number for this SprintLink address block.

Whenever an Internet datagram has a destination IP address whose first 15 bits match this pattern, it will be forwarded into the SprintLink network. Once it has reached the SprintLink network, bits in the destination IP address will be compared with longer network prefixes that match addresses assigned to SprintLink customers. The best (longest) match is the one that is chosen. For example, datagrams with destination addresses that start with 11001110 10011111 00 will be forwarded to Wolfe Internet access.

Most people do not like to work with bit patterns, so a couple of other ways of writing down network prefix information have been invented. For example, the notation below says: "Write out this address in binary, and then use the first 18 bits as the network number."

206.159.0.0/18

Of course, you let computers perform this conversion for you automatically so you never have to look at the bits.

Another (perhaps clumsier) way to describe the length of a prefix is to use a *network mask*. This is a quantity with 1s in the network part of an address and 0s elsewhere. The mask corresponding to the example above has 18 1s and is:

11111111   11111111   11000000   00000000

Converted to decimal, this is 255.255.192.0. The combination:

206.159.0.0, network mask 255.255.192.0

is another way of saying "Use the first 18 bits as the network number."

# Implications of Using ISP Addresses

If, like many organizations today, you obtained your Internet addresses from your ISP, then changing ISPs can be a very painful experience. This is because you must give up your current Internet addresses unless you can come up with an extremely compelling reason to keep them.

Changing addresses is necessary in order to preserve efficient routing. For example, suppose that Automated Data Systems wanted to change ISPs, but insisted that it had to keep its Class C network number, 206.158.2. The path to this specific network number would have to appear in every Internet backbone routing table, since Automated Data Systems could no longer be reached via the more general SprintLink route entry. If enough companies were allowed to walk away with their addresses, the entire hierarchical routing system would break down.

Some organizations try to evade this problem by obtaining Internet addresses directly from a registry. These addresses would not be tied to any ISP. However, if you do not use addresses from your own ISP's address space, you will discover that unless your company is very big and has plenty of clout, the ISP community may simply refuse to route your traffic. The pressure to use addresses from an ISP number space is very strong.

Fortunately, there are some strategies that you can follow to minimize the pain of address renumbering. If you have a large network, the most effective strategy is to use *private enterprise addresses* (which are described in the next section) in the bulk of your network. On the other hand, if you are connecting a single LAN to the Internet, then you can use dynamic address assignment for your desktop systems. This will make a changeover simple. We'll describe these strategies in the sections that follow.

# Private Enterprise Addresses

The Internet Assigned Numbers Authority has reserved the blocks of addresses shown in Table 7.4 for private use. These addresses never will be visible on the Internet. Any organization is free to use as many of these addresses as it wishes. These addresses can be used for:

- Systems that will not communicate with the Internet.
- Systems that can communicate with the Internet in a satisfactory manner via an intermediate system—an application proxy server or a network address translation server.

Recall that a proxy server interacts with Internet application servers on your behalf. A network address translation server translates internal addresses to one or more real Internet addresses.

**TABLE 7.4**

Private Enterprise Addresses

| Class | Private Address Blocks |
|-------|------------------------|
| A | 10.0.0.0 to 10.255.255.255 |
| B | 172.16.0.0 to 172.31.255.255 |
| C | 192.168.0.0 to 192.168.255.255 |

# Hiding Addresses via a Proxy Firewall

We described how a proxy firewall could be used to connect a site to the Internet in Chapter 5. Figure 7.2 illustrates how World Wide Web access works via a proxy. When a user requests a Web page, the user's browser opens a connection (1) to the proxy firewall, and sends the user's request to the server. In Figure 7.2, the proxy firewall is connected to a LAN that has private addresses. It is also connected to the Internet and has a real Internet address. The proxy opens a connection (2) to the Internet Web server and obtains the desired page. The proxy then sends the page back to the user via connection (1). Figure 7.3 shows how a proxy firewall on a perimeter LAN that has public Internet addresses can be used to provide shielded Internet access to users within the private part of an organization.

A proxy does more than hide your internal IP addresses. Using a proxy server has the benefit of adding substantial security defenses to your site. Internal computers are protected from Internet hackers because the internal systems never communicate directly with external computers.

Figure 7.4 shows how an end user can configure a Netscape World Wide Web browser to work via a proxy. If **Manual Proxy Configuration** is chosen, the user will be prompted to:

- Identify the proxy to be used for each application (World Wide Web, file transfer, etc.).

**Figure 7.2**
LAN computers accessing the Internet via a proxy firewall.

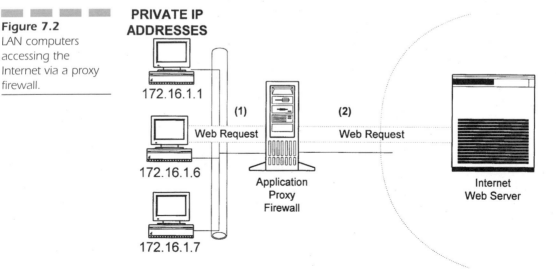

PRIVATE IP ADDRESSES

172.16.1.1

(1)

Web Request

172.16.1.6

172.16.1.7

Application Proxy Firewall

(2)

Web Request

Internet Web Server

**Figure 7.3**
Private site
computers accessing
the Internet via a
proxy.

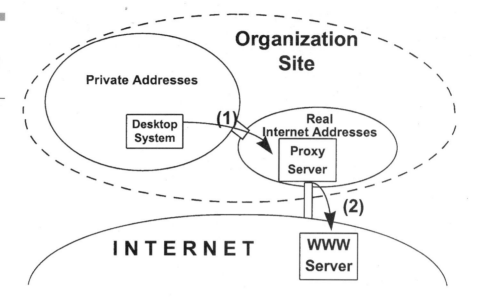

**Figure 7.4**
Configuring a
Netscape browser to
communicate via a
proxy.

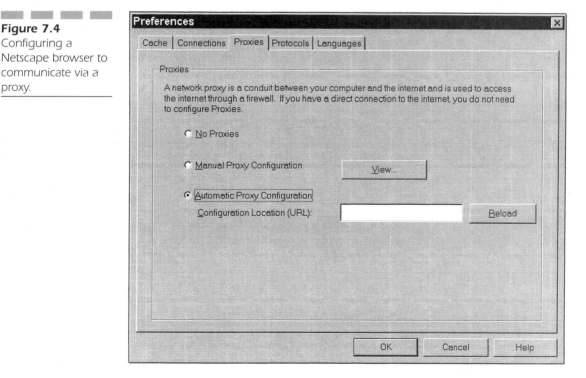

■ List hosts on the organization's own network that should be accessed directly (without relaying requests to a proxy system).

A network administrator can place this information into a central configuration file so that users can choose **Automatic Proxy Configuration** and simply enter the URL that identifies the file.

After proxy configuration has been completed, users operate their browsers in a completely normal way and can forget about the existence of the proxy system.

Proxy servers perform quite well. However, they operate at the application level, have a lot of connections to maintain, and have to keep a lot of information in memory. A good proxy certainly has no trouble keeping up with transmission speed when a site is connected to the Internet via one or two T1 lines.

# Network Address Translation Server

The network address translation (NAT) approach provides a lower level of security, but supports high transaction rates.

The basic idea is that a NAT server is given a pool of Internet addresses. The NAT server assigns these addresses temporarily on demand. That is, it translates your internal private address to a valid Internet address for outgoing traffic, and translates the Internet address back to your address for incoming traffic. Figure 7.5 shows an example in which internal private source address 172.16.1.50 is translated to 198.100.100.4 for datagrams in a session with an Internet World Wide Web server.

Note that with a NAT solution, internal computers *do* communicate directly with Internet computers. The basic NAT function is just a mechanical address translation.[3] However, some NAT products provide value-added security features such as:

■ Screening out undesirable Internet traffic

■ Enabling you to assign specific Internet access rights to your internal users

---

[3]NAT's job is just address translation, but doing this right is tricky. For example, it is commonplace for a file transfer client to send its address to a file transfer server as application data. A NAT box has to watch for this and substitute the public address. If you use NAT, be sure to choose a reliable, well-established vendor.

**Figure 7.5**
Network address
translation.

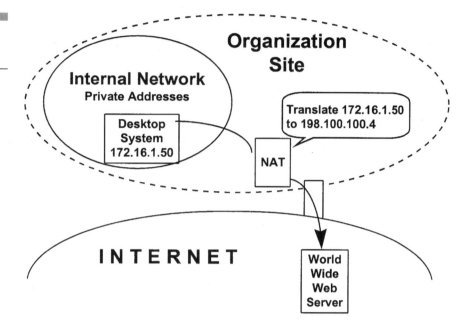

- Enabling you to control which of your internal computers may be
  reached from the Internet

When using a proxy or a NAT box, systems on your private network must
be assigned addresses that will not be visible somewhere on the Internet. Obviously, you can use any of the reserved private addresses. Over the past few
years, quite a few organizations that were unconnected to the Internet registered and obtained unique address blocks and assigned these addresses to
their computers. If you have unique addresses, you can continue to use them
privately if you wish. Just obtain some public Class C addresses that will be
visible to the outside world and use these on an intermediate LAN or in a
NAT box.

Some NAT servers translate all internal IP addresses to a single Internet
address. They make the sessions originating at internal users look like sessions originating at different clients logged into a single computer.

# Using Dynamic Host Configuration

One way to avoid the misery of reconfiguring your desktop TCP/IP systems
(for example, when you change ISPs) is to use a Dynamic Host Configuration

Protocol (DHCP) server. You assign IP addresses to desktops by entering a block of addresses (such as 198.207.177.1–198.207.177.62) at the DHCP server. When a desktop system boots, the server assigns the system an IP address that belongs to this block.

Free DHCP server software is available for Unix systems and is packaged with NT Server.

# Connecting NetWare LANs to the Internet

There are several NetWare Internet gateway products[4] that work in much the same way as a TCP/IP NAT server combined with an application proxy. These products enable NetWare users to access the Internet without installing TCP/IP at their desktops. As we noted earlier, gateway software at a NetWare server translates between NetWare's IPX/SPX communications protocols and TCP/IP. Users can run applications (such as Netscape) at their desktops.

A NetWare gateway server can assign a separate IP address to each client or can use a single IP address. This makes sessions originating at internal users look like sessions originating at different clients logged into a single computer.

# Subnets

If your network consists of a single LAN, then you can use IP addresses that have a very simple structure: a network number and a host number, as shown in Figure 7.6.

Sites with more complex networks need to use three-part addresses. A host's IP address must contain the information needed to route datagrams to

---

[4]For example, NOV*IX from FTP Software or Instant Internet from Performance Technology.

**Figure 7.6**
Simple IP address structure.

| NETWORK NUMBER | Host Number |
|---|---|

a system. To do this, it identifies the destination network, the LAN (subnet) to which the host is attached, and finally, the specific host on that subnet. To do this, a site's IP addresses are broken into three components: the network number, a subnet number, and a host number.

Today, 32-bit IP addresses are broken into three components on any convenient bit boundaries. For simplicity, we'll start our discussion with some examples based on traditional Class A, B, and C addresses.

A large organization that has a Class A or Class B network number would have many LANs. Even a small organization that has a single Class C network number might have a network made up of several small LANs. Figure 7.7 shows addresses that contain subnet numbers.

In Figure 7.7, the local part of each address appears to have been split into two equal parts in order to number subnets and hosts. In fact, your organization must choose where you want a split to occur. You can choose any convenient number of bits for your subnet field. Your choice depends on the maximum number of IP systems you wish to attach to any of your LANs. There are a couple of rules that you have to follow:

- Any field must contain at least 2 bits.
- The bits in a host address cannot all be 0s or 1s.
- The bits in a subnet address cannot all be 1s. All-0 subnets used to be forbidden too, but now are used at many sites.

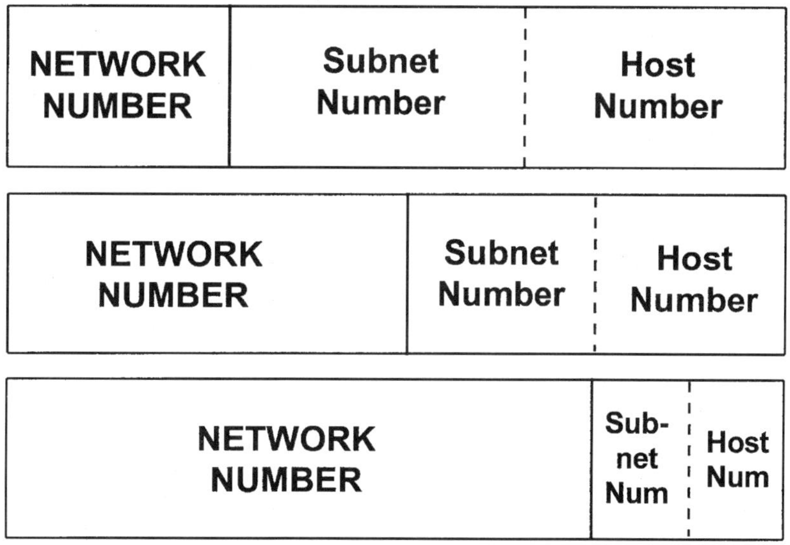

**Figure 7.7**
Addresses that contain subnet numbers.

Table 7.5 shows breakdowns for Class B and Class C networks. For example, if two bits are used to number subnets, you can use subnet numbers 0 (00), 1 (01), and 2 (10). Some of the values listed in the table are impractical today. Current LAN technologies do not support thousands of hosts.

If you have some very large LANs and some very small LANs, you can use different sizes for different parts of your network. If you need new addresses for a few LANs, you might prefer to get a separate Class C network number

**TABLE 7.5**

Numbers of Hosts and Subnets

| Number of Subnet Bits | Number of Subnets | Number of Host Bits | Maximum Number of Hosts per Subnet |
|:---:|:---:|:---:|:---:|
| CLASS B | | | |
| 2 | 3 | 14 | 16382 |
| 3 | 7 | 13 | 8190 |
| 4 | 15 | 12 | 4094 |
| 5 | 31 | 11 | 2046 |
| 6 | 63 | 10 | 1022 |
| 7 | 127 | 9 | 510 |
| 8 | 255 | 8 | 254 |
| 9 | 511 | 7 | 126 |
| 10 | 1023 | 6 | 62 |
| 11 | 2047 | 5 | 30 |
| 12 | 4095 | 4 | 14 |
| 13 | 8191 | 3 | 6 |
| 14 | 16383 | 2 | 2 |
| CLASS C | | | |
| 2 | 3 | 6 | 62 |
| 3 | 7 | 5 | 30 |
| 4 | 15 | 4 | 14 |
| 5 | 31 | 3 | 6 |
| 6 | 63 | 2 | 2 |

for each LAN. However, if each LAN contains a small number of systems, your ISP will tell you to subnet one or two Class Cs.

Many people dislike working with addresses that have been broken up along odd bit boundaries. They prefer to use byte boundaries so that they can look at an address such as 130.132.4.56 and say "This corresponds to host 56 on subnet 4 of network 130.132." Remember that you can use as many private addresses as you like inside your internal network, and avoid fiddling with bits.

You probably will need to learn the art of subnetting if you want to allow systems located on multiple LANs to communicate directly with the Internet (i.e., not through an application proxy or network address translator).

# A Subnetting Example

It is possible to learn to live with subnetting along odd boundaries, even though you may never learn to love it. Let's look at a fairly complicated subnetting example. Suppose that you have been assigned Class C network number 198.100.100 and you have four LANs. You project that over the next three years:

LAN A will have at most 60 systems

LAN B will have at most 50 systems

LAN C will have at most 20 systems

LAN D will have at most 20 systems

Table 7.6 shows how you could break up a single Class C address into 4 subnets that meet your needs. Note that two different subnet sizes are used.

**TABLE 7.6**

Breaking a Class C Network into Subnets

| LAN | Subnet Prefix (binary) | Number of Usable Host Addresses | Host Address Range |
|-----|-----|-----|-----|
| A | 00 | 62 | 198.100.100.1–198.100.100.62 |
| B | 01 | 62 | 198.100.100.65–198.100.100.126 |
| C | 100 | 30 | 198.100.100.129–198.100.100.158 |
| D | 101 | 30 | 198.100.100.161–198.100.100.190 |

**Figure 7.8**
Using one Class C
address for four
LANs.

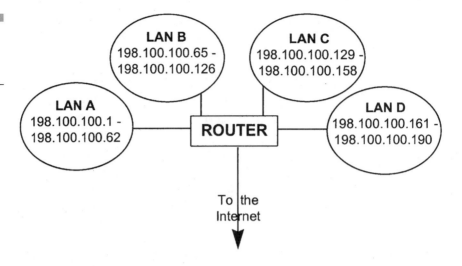

Figure 7.8 shows a site that has been broken into these four LANs. There still is a remaining block of addresses that can be used to add new LANs in the future.

## Subnet Masks

Earlier we described network masks, which are used to express how many bits are being used in the network part of an address. *Subnet masks* are used to express how many bits are used in the subnet part of the address. A subnet mask has 1s in the network and subnet parts of an address and 0s in the host part. (In other words, there are 1s everywhere except in the host part.) These masks are converted to dotted decimal notation or hexadecimal notation to make them easy to read and write.

Two subnet bits were used for LANs A and B in the previous section. These bits are shown in bold in the subnet mask below:

| 11111111 | 11111111 | 11111111 | **11**000000 |
|----------|----------|----------|--------------|
| 255 | . 255 | . 255 | . 192 |

Three subnet bits were used for LANs C and D. These bits are shown in bold in the subnet mask below:

| 11111111 | 11111111 | 11111111 | **111**00000 |
|----------|----------|----------|--------------|
| 255 | . 255 | . 255 | . 224 |

## Importance of Subnet Masks

Subnet masks help IP to route traffic. When an Internet datagram with destination address 198.100.100.82 arrives at the site router in Figure 7.8, the router has to figure out whether to deliver the datagram to LAN A, B, C, or D. The router uses the subnet mask corresponding to each router interface to block out the host bits. If the remaining address bits match, then the router has found the target LAN.

If the router thought like a person, it would look up the address in a table like Table 7.6. But routers can get the right answer very quickly by performing a bit-oriented computation using the subnet mask.

## IP Addresses Are Interface Addresses

We have looked at addresses at the network and subnet level. Now we are ready to assign IP addresses to systems. However, because of the way that IP addresses are structured, they are not assigned to systems, but to interfaces.

The router in Figure 7.9 is connected to two Ethernet LANs. This router is attached to LAN 16 on network 128.1., therefore the interface to this LAN is assigned an IP address that starts with 128.1.16. This router also is attached to LAN 5 on network 128.1., therefore the interface to this LAN is assigned an IP address that starts with 128.1.5. Each IP address identifies the network and subnet to which an interface is connected.

**Figure 7.9**
Interface addresses
for a router.

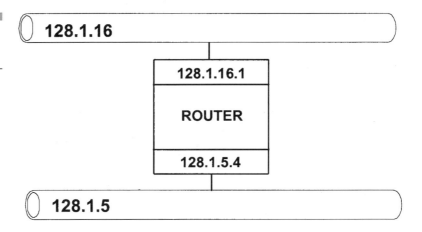

Below, we use the Windows NT *ipconfig* command to display interface configuration information for an NT server named *toshnt* that is attached to an Ethernet LAN (198.207.177) and a wide area point-to-point link (205.197.36). It is a peculiarity of the NT command output that it describes the wide area line as "Ethernet adapter NdisWan 5."

```
C:\> ipconfig

Windows NT IP Configuration

Ethernet adapter Elnk31:

        IP Address. . . . . . . . . : 198.207.177.41
        Subnet Mask . . . . . . . . : 255.255.255.0

Ethernet adapter NdisWan5:

        IP Address. . . . . . . . . : 205.197.36.52
        Subnet Mask . . . . . . . . : 255.255.255.0
```

An *ifconfig* command is used to obtain similar interface configuration information for a Unix host.

## Special Addresses

Some IP addresses have been set aside for special uses. These include:

- The loopback address, 127.0.0.1
- Broadcast addresses
- Zero host address
- Class D (Multicast) addresses

These are described in the sections that follow.

### Loopback Address

One of the nice features of TCP/IP is that it is very easy to test network applications, even when you do not have a network. You can run a client and a server inside the same computer. They communicate using the loopback address, 127.0.0.1.[5] By long-established tradition, this address corresponds to the name *localhost*.

---

[5]All addresses starting with 127 were reserved for loopback, but only 127.0.0.1 is used.

**TABLE 7.7**

Types of Broadcasts

| 255.255.255.255 | Broadcast to all systems on the local LAN. Used, for example, to request configuration information from a server at boot time. |
|---|---|
| 122.40.6.255 | Broadcasts a message to all systems on subnet 6. This might be sent by an administrator who wants to make an announcement. |
| 122.40.255.255 | Broadcasts a message to all systems on an entire Class B network. |

Below we see a command dialog between the text-based file transfer client-at NT Server *toshnt* and a file transfer server located at the same system. Note that 127.0.0.1 has been entered as the destination host.

```
ftp> open 127.0.0.1
Connected to 127.0.0.1.
220 toshnt Microsoft FTP Service (Version 2.0).
User (127.0.0.1:(none)): ftp
331 Anonymous access allowed, send identity (e-mail name) as password.
Password:
230 Anonymous user logged in.
ftp> dir
200 PORT command successful.
150 Opening ASCII mode data connection for /bin/ls.
dr-xr-xr-x 1 owner group   0 Jan 4 17:37 Disk
-r-xr-xr-x 1 owner group 1464 Jan 7 16:09 hosts.txt
dr-xr-xr-x 1 owner group   0 Dec 16 1996 info
```

A very good use of the loopback address is to test drive your own Web server using a browser client at the same system before putting the Web server online.

## Broadcast Addresses

A single message can be sent to all of the computers on a LAN or on an entire network by using a broadcast IP address. Table 7.7 presents examples of broadcast addresses. We assume that the third byte is used to number subnets for the sample network in the table.

*Security Alert:* Some prankster on the Internet might decide that it would be fun to send broadcasts to your network. You should configure your Internet router(s) to drop all incoming broadcast messages.

## All Zeros Address

An all-0s field cannot be used as a host address. A number like 122.40.6.0 is used to represent "subnet 6 on network 122.40" rather than to identify a specific host.

## Multicast Addresses

We have omitted a whole class of addresses from our discussion. Addresses whose first number is in the range 224 to 239 are called *multicast* addresses. These addresses are used for conferencing applications.

Users who wish to hold a conference need to obtain conferencing software. A conference organizer must choose an unused multicast address and the participants must configure their desktop conferencing software to initialize the use of this address. The conferencing software will carry out two steps:

- It will reconfigure the network interface card so that it will accept datagrams addressed to the multicast address as well as to its normal "unicast" address.

- It will notify the local router that this computer wishes to receive datagrams for the selected multicast address.

Figure 6.2 illustrated how information is sent to a multicast address. Note that when participants are spread across multiple networks, data must be forwarded by intermediate routers that have special multicast routing software. This is a fairly new technology, and this capability is just being phased in.

# REFERENCES

RFC 2050 *Internet Registry IP Allocation Guidelines,* K. Hubbard, M. Kosters, D. Conrad, D. Karrenberg, and J. Postel, November 1996.

RFC 1918 *Address Allocation for Private Internets,* Y. Rekhter, B. Moskowitz, D. Karrenberg, and G. de Groot, February 1996.

RFC 1631 *The IP Network Address Translator (NAT),* P. Francis and K. Egevang, May 1994.

RFC 1519 *Classless Inter-Domain Routing (CIDR): An Address Assignment and Aggregation Strategy,* V. Fuller, T. Li, J. Yu, and K. Varadhan, September 1993.

RFC 1518 *An Architecture for IP Address Allocation with CIDR,* Y. Rekhter and T. Li, September 1993.

# CHAPTER 8

# What You Need to Know about TCP/IP

In this chapter, we are going to describe the TCP/IP networking environment and TCP/IP concepts and terminology. We'll also look under the hood and examine some TCP/IP internals. Our goal is to provide a basic understanding of how things work so that you have information that you need in order to set up systems that enjoy good performance and are protected from attacks by network vandals. At the end of the chapter, we'll look at some examples that show how you configure systems for TCP/IP.

# Communications Protocols

TCP/IP is a family of *communications protocols*. A communications protocol:

- Is a set of rules that are followed when computers exchange information.
- Defines what each partner will do.
- Describes the format of the data that is passed between the partners.

# Hosts

*Hosts* are the sources and destinations for TCP/IP traffic. A host might be as small as a hand-held or notebook computer or as large as a mainframe. Any host in a TCP/IP network has the potential to communicate with any other host. That communication is based on client/server interactions.

When you use a desktop or laptop computer to send electronic mail, your computer is acting as an electronic mail client. When you visit World Wide Web sites, your computer is acting as a World Wide Web client. An important feature of TCP/IP is that any TCP/IP host can act as a client *and* as a server. For example, you can run a World Wide Web server on a Windows 3.1, Windows 95, or Macintosh computer. Of course, if you want to operate a high-performance server, you would use an NT Server with lots of memory (and maybe some extra processors) or a high-end Unix system.

# Routers

As we have seen, large TCP/IP networks are glued together by routers, which are special-purpose computers that forward data towards its destination. Figure 8.1 shows a personal computer named *mike.abc-corp.com* communicating with Web server *localwww.abc-corp.com*. The personal computer and Web server are on the same local area network (LAN) so they can exchange data directly. In contrast, the host *www.xyz.com* is located at a remote site that is reached by crossing the Internet. The data that *mike.abc-corp.com* exchanges with *www.xyz.com* passes through several routers.

The router that connects your site to the Internet is your doorway to the Internet. Configuring that router to screen out undesirable traffic is a very important chore. We'll introduce several screening criteria in this chapter. More details will be provided in Chapter 9 and Chapter 10.

**Figure 8.1**
Hosts communicating
using TCP/IP.

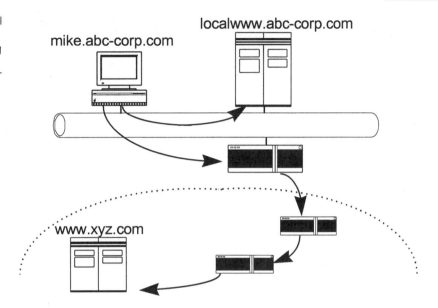

## Links and Subnetworks

A *link* or *subnetwork* is an Ethernet LAN, Token-Ring LAN, point-to-point line, frame relay circuit, or any other type of local area network or wide area connection. Routers tie different types of links together. The router in Figure 8.2 connects an Ethernet link and a Token-Ring link to a point-to-point link.

**Figure 8.2**
Links connected by a
router.

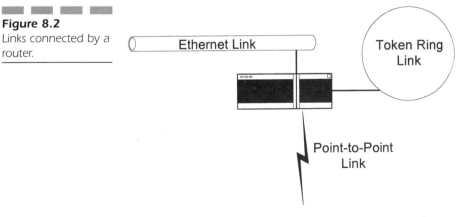

**Figure 8.3**
Forwarding a
datagram.

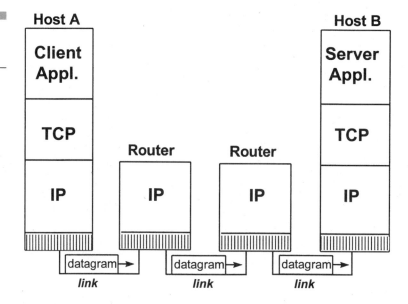

# IP Datagrams

IP, the Internet Protocol, defines the procedures that hosts and routers use in order to transport data from a source to a destination. Data is delivered in chunks that are called datagrams. Figure 8.3 shows a datagram that has been sent from a client application to a server application being forwarded from Host A to Host B across a sequence of links. IP is a very simple data delivery method.

# ICMP

IP is what is known as a "best effort" protocol. Hosts and routers will do their best to deliver datagrams, but sometimes they will fail. A link might be damaged, a router might be out of service, or the destination host might not be available.

When things go wrong, the source host is notified of the problem by a helper protocol called the *Internet Control Message Protocol,* or ICMP. Router X in Figure 8.4 knows that an adjacent link is broken and has sent an ICMP message back to Host A. The ICMP message notifies Host A that the destination cannot be reached.

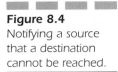

**Figure 8.4**
Notifying a source that a destination cannot be reached.

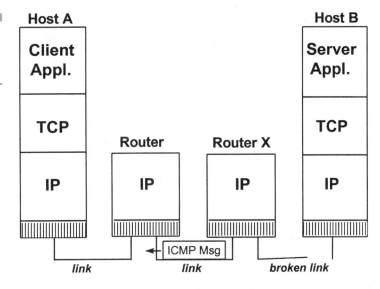

# TCP

The *Transmission Control Protocol,* TCP, provides the reliable sessions that are used to retrieve a World Wide Web page, copy a file, or deliver electronic mail. TCP builds on IP's rough forwarding service, making sure that data is delivered to the recipient application reliably and free of errors. As Figure 8.5 shows, TCP session data can flow in both directions at the same time.

**Figure 8.5**
TCP two-way session flow.

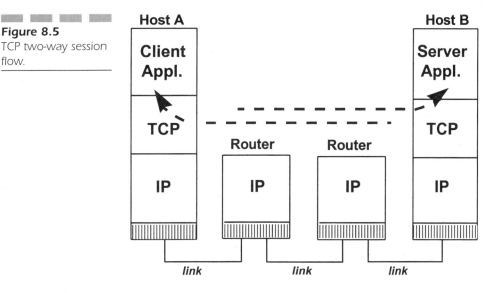

**Figure 8.6**
A UDP request and
response.

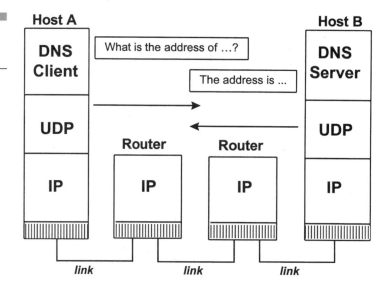

# UDP

The *User Datagram Protocol,* UDP, is another service that is built on top of IP. UDP is a very simple message delivery service that is just right for building simple request/response applications. We encountered a typical UDP application—Domain Name System queries—in Chapter 4. In Figure 8.6, a domain name query is sent to a DNS server, and a response containing the corresponding address is sent back. The request and the response are packaged inside UDP messages.

# TCP/IP Protocol Stack

By now, we have talked about several components of TCP/IP networking. Figure 8.7 shows how they all fit together. The components are piled on top of one another, and the whole layered configuration is called a *protocol stack.*

Years ago, the International Standards Organization (ISO) assigned names to each layer. These are shown at the left. Note that ICMP actually is considered to be part of the IP layer.

**Figure 8.7**
The TCP/IP protocol stack.

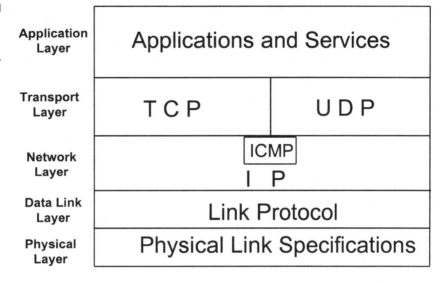

## Packaging Data

Before application data can be delivered, some extra information is added by each layer. Figure 8.8 shows how application data is packaged for transmission by adding this information.

**Figure 8.8**
Packaging data for transmission.

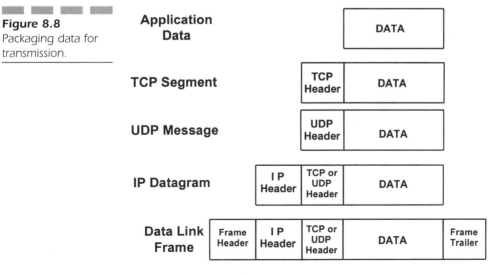

## TCP Segment

An application that uses reliable sessions (such as a file transfer) passes data to TCP. TCP adds a header, forming a *segment*. Among other things, each TCP header includes numbering that helps to keep data in its proper order.

## UDP Message

An application based on the simple exchange of messages (such as Domain Name System queries) passes a message to UDP. UDP adds a header, forming a UDP message.

## IP Datagram

TCP and UDP pass their segments and messages to IP for delivery. IP adds a header, forming an *IP datagram*. Among other things, the IP header contains the IP addresses of the destination and source hosts.

## Link Layer Frames

IP passes each datagram to data link software. A link layer header is put in front of the datagram and a trailer[1] is added at the end, forming a link layer *frame*. The frame header contains information that is needed in order to deliver the frame across the link. For example, for a local area network, the frame header will contain numbers that identify the destination and source network interface cards.

## Encapsulation

Wrapping a frame header and trailer around a datagram is called *encapsulation*. The frame headers used for each technology look different. In fact, there are two different frame header formats used for Ethernet LANs, and there are

---

[1]The frame trailer contains a quantity called a *Frame Check Sequence*. This is the result of a calculation on the 0s and 1s in the rest of the frame. The value is recalculated at the receiving end. If the answers do not match, some of the bits in the frame have been damaged during transmission and the frame will be discarded.

several different formats for leased lines. When you set up the router that connects your site to the Internet, you will need to identify the encapsulation formats that will be used on the links that are connected to the router.

# Framing and Reframing

A frame is good for one hop across one link. To understand what this means, think of a frame as a ticket that a datagram needs in order to take a ride across a link. If the datagram has to cross several links, it will need a separate "ticket" for each link.

By analogy, suppose that you were going to visit an island, and to get there:

- First you had to buy a train ticket and take a train.
- Then you had to buy a bus ticket and ride on a bus.
- Finally, you had to buy a boat ticket and board a boat.

After the train hop, you might as well throw away your train ticket. It's been used, and won't get you anywhere any more. Next you will use up the bus ticket, and then the boat ticket.

Frame encapsulations work the same way. After the trip across a link is complete, the frame header and trailer are useless. They are stripped off and thrown away.

# How IP Routes

When a LAN computer needs to transmit a datagram, IP checks whether the destination is on the same LAN. If it is, the datagram is wrapped in a frame and sent straight to the destination. If the destination is remote, then the datagram is wrapped in a frame and sent to a router connected to the LAN. Usually, systems are configured with the address of a default router, and all remote traffic is forwarded to this router.[2]

A router forwards a datagram by looking up its destination in its routing table. Routing table entries can be entered manually by an administrator, or else are learned from other routers via periodic exchanges of information.

---

[2]Sometimes more than one router is attached to a LAN. In this case, the default router will inform local hosts of exactly which traffic should be sent to a different router.

**Figure 8.9**
Finding the neighbor
who has a given IP
address.

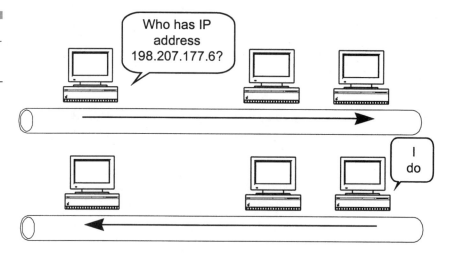

## LAN Address Resolution

When you send a frame across a LAN, the frame's header has to identify the destination's network interface card number. How does IP find this out? The answer is that it asks, using a procedure called the *Address Resolution Protocol*. As shown in Figure 8.9, a broadcast is sent across the LAN saying "Who has this IP address?" The target system replies, providing its network interface card number.

## CIDR Routing

Unlike the telephone system's country codes and area codes, Internet addresses originally had no structure that would help to locate destinations. For example, network 128.36 was assigned to Yale University in New Haven, Connecticut and network 128.37 was assigned to the Army Yuma Proving Ground in Yuma, Arizona. The result was that the routers that deliver information across the Internet had to store huge amounts of information in order to find paths to the thousands of individual network destinations.

However, the new hierarchical numbering structure described in Chapter 7 simplifies routing for more recently registered sites. For example, a single Internet routing table entry can now provide the information that SprintLink is responsible for all network numbers in the range 206.158.0–206.159.255. Of course, routers inside SprintLink's network need to maintain detailed information on

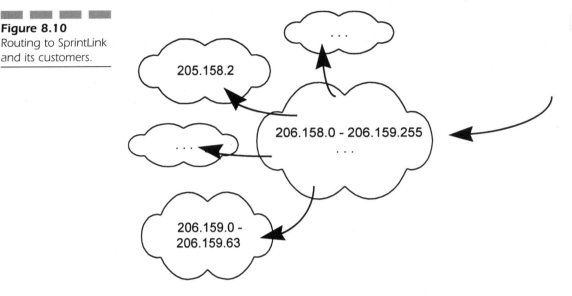

**Figure 8.10**
Routing to SprintLink
and its customers.

how to route to each of SprintLink's own customers. Figure 8.10 illustrates how the outside world can view SprintLink and all of its customers.

The hierarchical routing procedure that was made possible by the new bit-oriented classless way of handing out blocks of addresses is called *Classless Inter-Domain Routing* or CIDR.

# IP Header

The most important fields in an IP header are displayed in large bold print in Figure 8.11. These include:

■ The destination IP address.

■ The source IP address.

■ The protocol field. The recipient IP will read the protocol field. If the field value is 6, the datagram contents will be handed to TCP. If the value is 17, the datagram contents will be handed to UDP.[3]

Destination and source address fields are examined by screening routers to determine whether a datagram should be forwarded. For example, an

---

[3]As you might guess from the numbers, there are other possibilities. For example, if the datagram is carrying an ICMP error message, the protocol number is 1.

application proxy and a Web server may be the only systems to which incoming traffic may be addressed.

The protocol field also is used for screening. Hackers have been known to attack a site by streaming UDP datagrams at the site, clogging the local network with junk. Screening routers or firewalls can be configured to scrutinize datagrams whose protocol value is 17 very carefully. Many sites block out all incoming UDP traffic except for responses to Domain Name System queries.

# Fragmentation

We've seen that a datagram must be encapsulated in a frame before it can be sent across a link. One of the facts of networking life is that every type of link technology imposes a different size limit on how big a frame can be. For example, an Ethernet frame consists of at most 1518 bytes, while on a 16-megabit-per-second Token-Ring, frames may be bigger than 17,000 bytes.

What happens when a large datagram needs to be forwarded across a link that can only carry small frames? An IP router can chop a big datagram into several smaller datagrams. The destination host puts the pieces back together again. The fields in the second row of the IP header enable the destination host to reassemble a fragmented datagram. The *identifier* field shows which pieces belong together, and the *fragment offset* field shows where each piece goes.

Fragmentation is very bad for performance. If a lot of fragmented traffic is sent or received across your Internet link, you will have sluggish response time. There are several methods used to prevent most fragmentation. Consult a more advanced text to find out the full story.

# Other IP Header Fields

The header in Figure 8.11 is used with the prevalent version of IP, which is version 4. IP next generation (which is version 6) has a different header format. For version 4:

- The usual header length is 20 octets. However, there are some options fields that occasionally are added, making the header longer. The use of options in ordinary commercial or educational traffic is extremely rare.

- Precedence (priority) values can be used to request that routers give preferential treatment to some traffic.

- The Type of Service value can be used to request that routers provide low delay, good throughput, or reliable delivery.

- The Time-To-Live field contains a hop counter that is used to check whether datagrams have been in the network too long. The value is decreased by one at each router. If the value reaches zero at a router, then the datagram is discarded.

- A checksum is computed against all of the fields in the IP header, and is stored in the checksum field. IP at the destination system recalculates the checksum and discards the datagram if the answer does not match the value in this field.

**Figure 8.12**
Congestion.

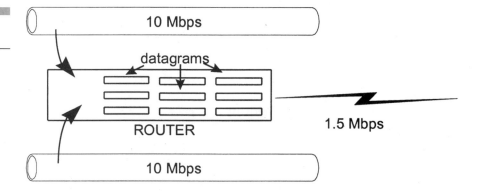

## IP Performance

IP performs well for several reasons:

■ Datagram processing is simple and can be done very quickly.

■ When a datagram is ready to be sent and link bandwidth is available, the datagram can be transmitted immediately.

Today's routers can run at "wire speed." This means that they can use all of the bandwidth that is available. However, IP cannot perform miracles. The top half of Figure 8.12 shows two 10-megabit-per-second Ethernet LANs at a site feeding traffic onto a 1.5-megabit-per-second T1 line that connects to the Internet.

If too many of your hosts are sending traffic to remote Internet hosts, then outgoing datagrams will pile up in the router's memory, waiting to be sent. Eventually, the router's memory will fill up and the router will have to throw datagrams away instead of delivering them. This situation is called congestion. (Fortunately, when congestion occurs, TCP automatically throttles back on the amount of data that it sends, and helps the router to recover.)

Congestion can occur at the router that connects your site to your ISP, at a router in your ISP's network, or at a router elsewhere on the Internet that is carrying your data to its destination.

## More about ICMP

There are several situations that will cause IP datagrams to be discarded. For example:

- There is a break in the network and the destination cannot be reached.
- A routing problem causes datagrams to wander around until the Time-To-Live hop counter expires.
- A router might be overloaded, forcing it to discard some datagrams.

ICMP messages are used to report these and other problems. An ICMP message also is the basis of the useful *ping* application, as well as messages that announce the presence of routers.

# ICMP and Security

Some sites configure the router interface that connects to the Internet so that it will throw away all incoming ICMP messages. This prevents a hacker from disrupting good sessions by sending falsified error reports to your network. Some sites also block all outgoing ICMP messages to prevent hackers from receiving information that might be useful.

Blocking these messages will not harm your communications with the outside world. However, you might want to allow one or two administrator systems to perform *pings* and *traceroutes*, because these are useful troubleshooting tools.[4]

Here is one more tip that will make you a good Internet citizen. Configure your Internet router so that it will not send *Source Quench* messages, which are used to report congestion. These reports can make congestion worse and are now considered bad practice.

# UDP and TCP Applications

The User Datagram Protocol, UDP, and the Transmission Control Protocol, TCP, send and receive data on behalf of applications.

UDP enables clients and servers to exchange messages. We already have seen that Domain Name System queries and responses are good examples of

---

[4]To do this, you would specify the IP addresses of systems that will be allowed to send the requests and receive the responses that are used by the *ping* function. These systems also would be allowed to receive Time-to-Live (hop count) expired ICMP messages, which are used to implement the *traceroute* program.

UDP messages. UDP also is useful for applications that need to broadcast information. For example, you might subscribe to a service that streams current stock information to every desktop on your LAN. A UDP broadcast is a feasible choice for delivering this information.

In contrast, TCP provides reliable sessions between clients and servers. TCP is used for World Wide Web sessions, electronic mail delivery, file transfers, terminal sessions, network news access, and more.

# Application Port Numbers

Earlier, in the section entitled *IP Header*, we saw that IP at a destination host decides whether to pass incoming data to TCP or UDP by checking the Protocol field in the header. If the value is 6, then the data is passed to TCP. If the value is 17, then the data is passed to UDP.

Similarly, once incoming data has been passed to TCP or UDP, they need to figure out how to deliver it to applications. This is done by assigning numbers called *application port numbers* (or simply *port numbers*) that reveal which application should receive the data.

The use of the word "port" was not a very good choice. It makes network people visualize hardware connectors on the back of network devices and it makes most other people think of ships in a harbor. For TCP or UDP, a port corresponds to a location in computer memory where data for a particular application is parked so that the application can pick it up.

Port numbers range from 1 to 65,535. TCP and UDP port numbers are independent of each other. As shown in Figure 8.13, one host application may be receiving UDP messages at UDP port 5000 while a completely different application is engaged in a TCP session at TCP port 5000.

# Well-known Ports

Small port numbers from 1 to 1023 are reserved for well-known server applications. These are called *well-known ports*. For example:

■ World Wide Web servers are accessed at TCP port 80.

■ A file transfer server can be accessed at TCP port 21.

■ Electronic mail is delivered at TCP port 25.

■ Queries to a Domain Name Server are delivered to UDP port 53.

**Figure 8.13**
UDP and TCP
application port
numbers.

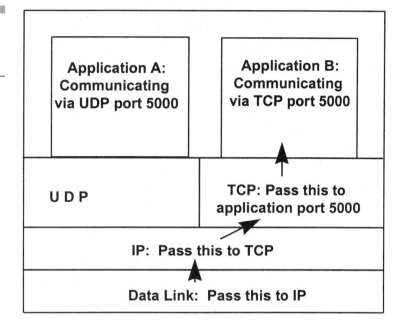

Your computer contains a file that lists well-known ports. On Unix systems this file is called */etc/services*. On Windows 95, the file is located at *C:\Windows\Services*. On Windows NT Server, it is called *C:\WINNT \system32\drivers\etc\Services*. The excerpt below was taken from the Windows 95 file (which has the same format as the NT file and Unix file).

```
ftp-data        20/tcp
ftp             21/tcp
telnet          23/tcp
smtp            25/tcp      mail
time            37/tcp      timserver
time            37/udp      timserver
rlp             39/udp      resource        # resource location
name            42/tcp      nameserver
name            42/udp      nameserver
whois           43/tcp      nicname
domain          53/tcp      nameserver      # name-domain server
domain          53/udp      nameserver
nameserver      53/tcp      domain          # name-domain server
nameserver      53/udp      domain
```

Well-known ports are assigned by the Internet Assigned Numbers Authority (IANA) and published in an RFC document called *Assigned Numbers*.

# Client Ports

The remaining port numbers (from 1024 to 65,535) are in a pool, and are assigned to client applications on an as-needed basis.[5] When a client no longer needs a port number, it goes back into the pool.

If you have a Windows 95, NT, or Unix system, you can check up on the port numbers that are being used in current, active sessions by typing the command: *netstat -na*. The display that follows was captured at a Windows 95 computer. To obtain the display, we opened a DOS command window, started Netscape and connected to *www.snet.net,* and then switched back to the command window and typed *netstat -na.* Note that connecting to the Web site caused a bunch of TCP sessions to be opened.

```
> netstat -na
Proto Local Address          Foreign Address        State
  TCP 130.132.57.25:1032     204.60.200.8:80        ESTABLISHED
  TCP 130.132.57.25:1033     204.60.200.8:80        ESTABLISHED
  TCP 130.132.57.25:1034     204.60.200.8:80        ESTABLISHED
  TCP 130.132.57.25:1035     204.60.200.8:80        ESTABLISHED
  TCP 130.132.57.25:1036     204.60.200.8:80        ESTABLISHED
  TCP 130.132.57.25:1037     204.60.200.8:80        ESTABLISHED
```

To reach the Web site, our browser client:

- Looked up the IP address of *www.snet.net,* which is 204.60.200.8.
- Obtained a client port, 1032, from the local Windows 95 TCP service.
- Connected to the World Wide Web port, 80, at 204.60.200.8.
- Retrieved the Web site's default home page.
- Opened separate TCP sessions (using client ports 1033, 1034, 1035, 1036, and 1037) to retrieve the images on the home page.

We will go into the details of the way that a Web browser and server interact in Chapter 13. By the way, Web TCP sessions come and go quickly, so you might miss some of them if you do not switch back to DOS immediately.

# Socket Addresses

The information that is needed to reach a remote application consists of its IP address and port number. The combination of an IP address and a port number is called a *socket address.*

---

[5]There are a few servers that run at higher-numbered ports, such as Network File Servers, which use port number 2049.

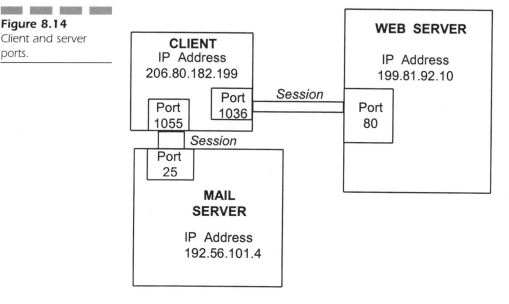

**Figure 8.14**
Client and server ports.

A computer can participate in many sessions at the same time. For example, you could access a Web site and retrieve your electronic mail at the same time. Each client session uses a different port number, and therefore it has a different socket address. However, as shown in Figure 8.14, an application service uses the *same* port number for all of its sessions.

# Understanding TCP Performance

TCP, the Transmission Control Protocol, provides reliable sessions. To do this, TCP:

- Sets up and terminates sessions between clients and servers.
- Makes sure that all the data that was sent is delivered safely, in order, and free of errors.

Over the years, TCP has been honed and refined to perform better and to assist in recovering from network congestion. Our goal in the next few sections is not to describe all of TCP's wonderful mechanisms. We wish to provide a sufficient understanding of the things that you will need to know in order to tune your clients and servers for good performance and configure security screening for TCP sessions.

# Transferring Data Reliably

We already know that TCP adds a header to a chunk of data and passes it to IP for delivery. But what does TCP do to make sure that the data is delivered reliably? Through all of the years that computers have communicated, there has been one basic method of delivering data reliably. That method is as follows:

- A number is placed into the header.
- A clock is set with a timeout value.
- The message is transmitted.
- If an *acknowledgment* of that data arrives from the partner, the data has been delivered safely.
- If the timeout expires and an acknowledgment has not arrived, the data is retransmitted.

There is a limit on the number of retransmissions. If this limit is reached and the data still has not been acknowledged, then the session is declared dead.

# Receive Window

As shown in Figure 8.15, TCP can send and receive data at the same time. Furthermore, lots of datagrams containing TCP segments can be in the pipeline at any moment. The amount of data that can be in the pipeline is limited by only one thing: the amount of memory space that the receiver currently

**Figure 8.15**
Pipelining TCP
segments.

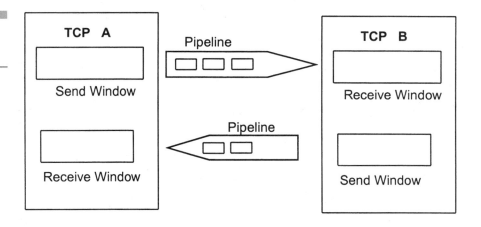

has available for storing incoming data for the session. This memory space is called the *receive window*. Every segment that TCP transmits contains an acknowledgment of received data and an update on the current size of the originator's receive window.

Data that is waiting to be sent is stored in a *send window*.

# Numbering Data

TCP numbers data in a somewhat unusual way. Each byte of data is numbered! When TCP packages a block of bytes into a segment, the number corresponding to the *first* byte in the segment is put into the TCP header. For example, Figure 8.16 shows three TCP segments. The first contains bytes 5001 through 6000, the second contains bytes 6001 through 6505, and the third contains bytes 6506 through 6896.

- The number corresponding to the first byte is placed into the outgoing TCP header.
- The number of the next byte that is expected is used for the acknowledgment.

By the way, at the beginning of a session, numbering does not start at 1. Each partner chooses a start number, and there is some randomness involved in the choice. This turns out to be important for session security, and helps to prevent vandals from hijacking sessions.

# Starting a TCP Session

A client and a server exchange some useful and important information when a session is started. When a client requests a connection:

- The client announces its starting point for numbering the data that it will send.

**Figure 8.16**
Numbering TCP
segments.

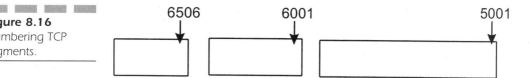

- The client also announces the size of its receive window.

- Optionally, the client can announce the size of the biggest incoming data segment that it can handle.

The server responds, announcing its start number, receive window size, and optionally, the size of the biggest incoming data segment that the server can handle.

This exchange of information is called the *opening handshake*.[6] After these announcements, each has to respect the partner's wishes. For example, the server must not send data in chunks that are too big, and the server must not send more than the client's window can hold.

## Performance Impact of Receive Window and Maximum Segment Size

A system's window size and maximum segment size have an enormous impact on performance. For example, a World Wide Web server sends a lot of data to a client. You will get the best performance if the client can receive big segments and has a big receive space. On the other hand, you should not use segments that are so big that they would cause fragmentation. Currently, 1460-byte segments work well, and usually a receive space that is eight or sixteen times this size is used.[7]

Unfortunately, there are browser implementations that do not announce any segment size during session start-up. When this happens, an Internet server will send data in 536-byte segments, which has a devastating effect on performance.

If you wish to get the best possible throughput out of your link to the Internet, make sure that you use client software that announces an appropriate maximum segment size and provides plenty of receive space. For example, Netscape and Microsoft browsers (and their Web applications) announce good maximum segment sizes and provide reasonable receive memory allocations.

---

[6]The full handshake consists of these messages plus an acknowledgment from the client.

[7]The largest size currently permitted is 65,536 bytes.

# Maximum Processes and Listen Queue

If you operate a server, two parameters may be choking down the number of clients that it can handle, even when there is plenty of bandwidth and plenty of extra capacity at the server. One of these parameters is a limit on the maximum number of concurrent clients that can be handled. A limit is needed—otherwise the server will run out of resources and crash—but the limit should not be set too low.

The other parameter is rather odd. During the opening handshake, a client is parked outside of a server on a *listen queue*. Many computers are configured so that by default the listen queue can only hold 5 clients. Additional clients are turned away, even when the server is nowhere near its maximum number of concurrent clients. A five-slot listen queue is much too small for a busy Web server, and needs to be changed.[8]

# Slow Start

What is the first thing that you do when you arrive at work? Many people start the day by reading their electronic mail. But a network can be pretty badly stressed when many people start to transfer data at the same time.

TCP's solution is called the *slow start*. At the start of a session, TCP will send one segment and wait to receive its acknowledgment. Across the Internet, this might take a half second or more. Then TCP can send two segments. If their acknowledgments come back safely, TCP can send four segments. This doubling goes on until TCP reaches the limit imposed by the receiver's window size.

When a network gets congested, TCP throttles back and performs another slow start. Slow starts can have a very visible impact on the delivery of a Web page when small segments are used. This provides another argument in favor of using products that announce a reasonable segment size.

---

[8]You might need to change an operating system parameter, and also change the *listen( )* program call in your server's source code and recompile, in order to fix this problem. Check your system and product documentation.

# Closing a TCP Session

Very often, it is the server that closes a TCP session. Typically, a user indicates that the work is over by typing "quit" or by clicking on "exit" on a menu. A message is sent to the server application at the application level. The server application then tells its TCP to start a close operation.

There are two features of the close process that can spell trouble for a Web server, so we will look at the procedure in detail. As shown in Figure 8.17, the closer sends a segment called a FIN to the partner. The partner replies with an acknowledgment (ACK). Then the partner sends its own FIN separately. The closer sends its acknowledgment and enters a waiting state that lasts a few minutes. Finally, the closer deletes the session.

The closer's waiting state is called TIME_WAIT, and often lasts for 4 minutes. During this time, server memory resources that were assigned to the session cannot be reclaimed and used by other tasks at the computer. This is particularly irksome at a Web server, where retrieving a single page may trigger 20 TCP sessions and therefore potentially cause 80 minutes of frozen resources.

Fortunately, some Web browsers respond to the server's request for a close with an abort message instead of terminating the session according to the rules. The server then can release the memory resources immediately. This is not elegant, but it does help the server to be more efficient. For example, browsers from Netscape and Microsoft follow this procedure.[9] Unfortunately, some browsers do not.

---

[9] The most recent version of the Web protocol, HTTP 1.1, will ease this problem. See Chapter 13.

**Figure 8.17**
Closing states.

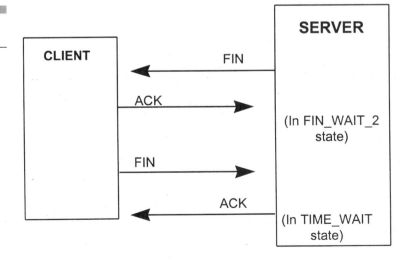

There is another potential problem that is even worse. Some clients never send their FIN message, and there currently are some Web server implementations that will sit and wait forever for the client to make its polite response.[10] These servers never release the memory that was reserved for these dead sessions. The result is that eventually all of the server's memory fills up and it crashes.

# Contents of the TCP Header

Figure 8.18 shows the format of a TCP segment. We will describe the TCP header fields that are of greatest interest.

---

[10] While a server is waiting for the FIN, it is said to be in FIN_WAIT_2 state.

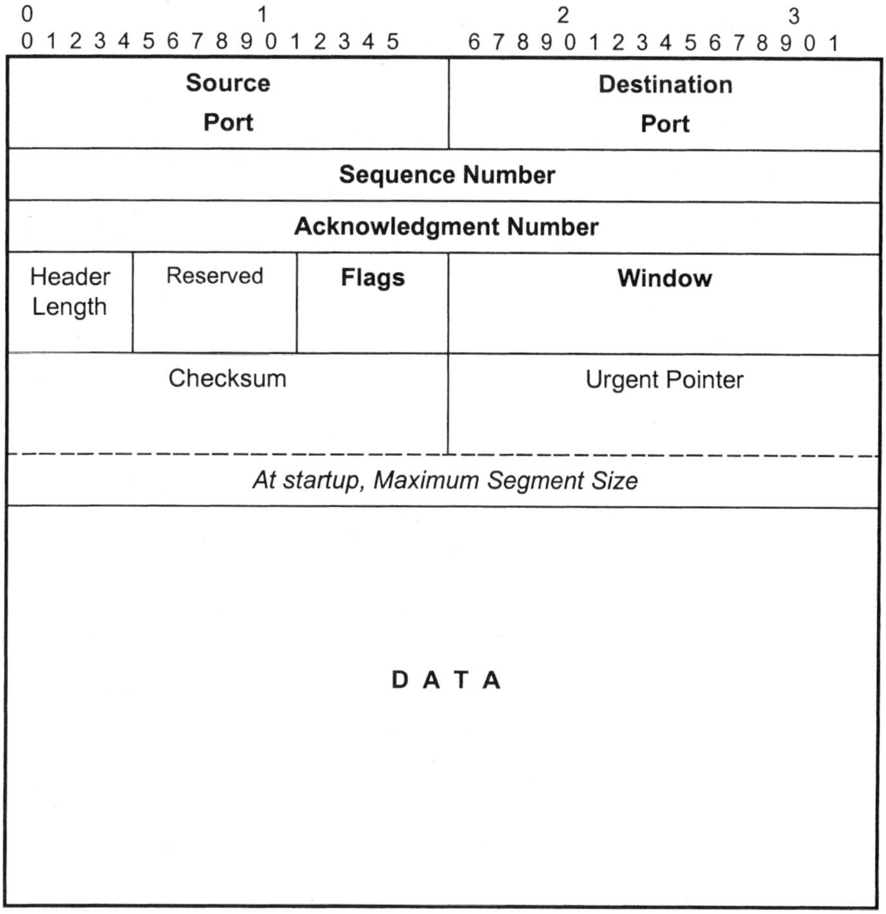

**Figure 8.18**
Format of a TCP
segment.

| 0 | 1 | 2 | 3 |
|---|---|---|---|
| 0 1 2 3 4 5 6 7 8 9 0 1 2 3 4 5 | | 6 7 8 9 0 1 2 3 4 5 6 7 8 9 0 1 | |

| Source Port | Destination Port |
|---|---|
| Sequence Number | |
| Acknowledgment Number | |

| Header Length | Reserved | Flags | Window |
|---|---|---|---|

| Checksum | Urgent Pointer |
|---|---|

*At startup, Maximum Segment Size*

**D A T A**

When setting up a new TCP session, flags in the TCP header indicate that this is a request for a brand new session.[11] Filtering routers and screening firewalls watch for new sessions and apply preconfigured rules to decide whether the session will be permitted. Another flag is used to negotiate a graceful close[12], and a third flag signals that the session should be aborted.[13]

During data transmission, the most important fields in the TCP header are those that report:

■ The source and destination port numbers.

■ A sequence number for enclosed data.

■ An acknowledgment of data received from the partner.

■ A report on how much memory space is available to receive more data from the partner.

Filtering routers and screening firewalls use port numbers to control:

■ The Internet applications that local clients are allowed to access.

■ The local server applications that can be used by Internet clients.

# Configuring TCP/IP

Configuring TCP/IP is surprisingly easy. Once a system's LAN adapter has been properly installed, all that you need to do to get a computer ready for TCP/IP communications is to configure:

■ The interface IP address

■ The interface subnet mask

■ The IP address of a default router

■ IP addresses for one or more Domain Name Servers

We'll look at configuration for three sample systems: Windows 95, Windows NT Server, and Unix.

For Windows 95, the configuration menus are displayed by selecting **Start/Control Panel/Network**, choosing the entry that binds TCP/IP to your LAN adapter, and clicking on **Properties**. Figure 8.19 shows the Windows 95

---

[11]The ACK and SYN flags are used. In a session request, ACK=0 and SYN=1.

[12]The FIN flag. FIN=1 is used to close.

[13]The reset, or RST flag. RST=1 is used to signal the abort.

**Figure 8.19**
Menus for
Windows 95.

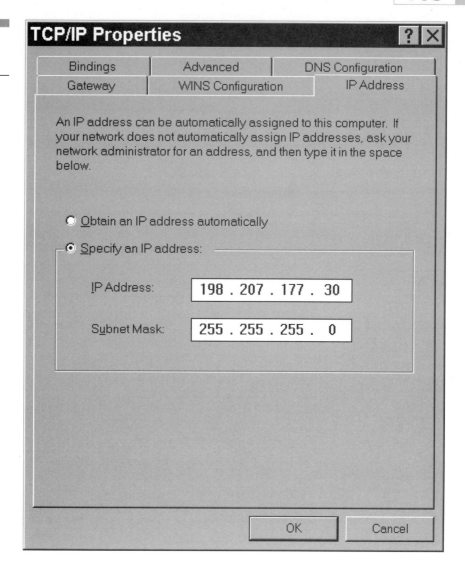

display (which we already viewed in Chapter 3). An IP address and subnet mask for the LAN interface have been entered into the IP Address menu.[14]

Configuration menus for Windows NT Server 4.0 are accessed via:

Start/Settings/Control Panel/Network

Figure 8.20 shows the resulting display.

---

[14] Systems also may be configured automatically by means of a Dynamic Host Configuration Protocol server.

**Figure 8.20**
Menus for Windows
NT Server 4.0
networking.

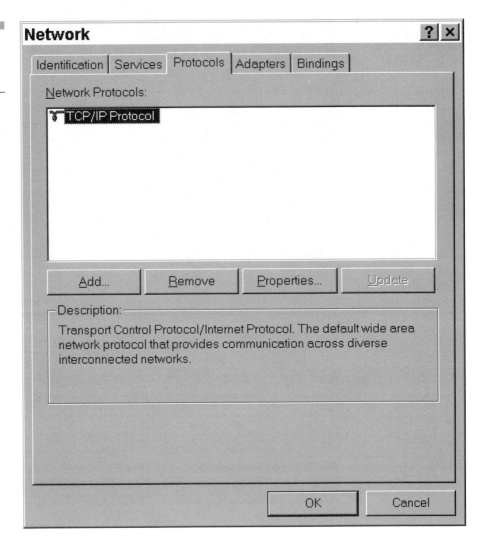

If you choose TCP/IP Protocol and Properties, you will see the configuration display in Figure 8.21. After filling in these menus, you will need to go back to the master menu, choose Services, and select all of the services that you wish to install and bind to TCP/IP communications.

Text-based commands are used to configure a Unix system. The *ifconfig* command is used to set, modify, or display the IP address and subnet mask

**Figure 8.21**
Configuring TCP/IP
for Windows NT
Server.

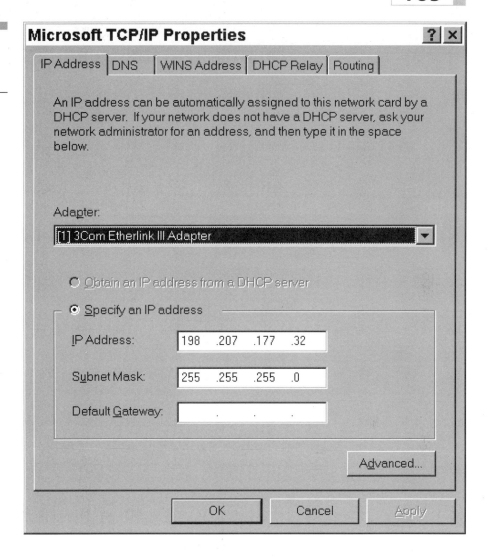

for an interface. Note that below, the subnet mask is written in hex instead of dotted decimal form.

```
ifconfig le0 inet 198.207.177.35 up netmask ffffff00
```

The *route* command is used to enter routing table entries. Below we define the default router.

```
route add default 198.207.177.1
```

Both *ifconfig* and *route* commands need to be written into a script that runs whenever the Unix system is booted.

The Unix configuration file */etc/resolv.conf* contains the name of the local domain and the addresses of Domain Name Servers. Below, an ordinary text-editing program (*vi*) is used to enter the information.

```
> vi /etc/resolv.conf
domain mgt.com
nameserver 198.207.177.41
nameserver 198.207.177.42
```

# Setting Up a LAN Interface Card

The previous section showed how easy it is to configure TCP/IP. The only problems in getting a desktop system networked arise because the LAN network adapter card was not installed successfully. Let's look at the issues involved in installing a LAN network adapter card. First of all, it is important to buy adapters that will make your job easy.

For Windows 3.X, be sure to get self-configuring cards. For Windows 95, make sure that the adapter has Plug 'n Play support. For Windows NT, make sure that you use an adapter that can be discovered and configured automatically by your version of NT.

Even when automatic adapter configuration is used, interrupt conflicts still occasionally occur and have to be fixed manually. Every device on your computer communicates with the computer's central processing unit (CPU) by means of an assigned interrupt number. Trouble can arise when two devices try to use the same interrupt.

You can save yourself some trouble by checking the documentation that comes with your adapter to find out what interrupts are usable for this card. Before installing the adapter, look over your current list of active interrupts and make sure that there is an interrupt that is both free and usable by the interface card. It is easy to check your list of interrupts for Windows 95. Select:

Start/Settings/Control Panel/System/Device Manager

Click on properties, and make sure that the interrupt request radio button has been selected.

Currently, you have to open several menus to find out which interrupts are in use for Windows NT. Once you are sure that you have a usable interrupt, install and configure the adapter card.

In Windows 95, go back to the Device Manager to make sure that your network adapter has been installed correctly. Click the "+" next to **Network Adapters** to view individual interfaces. If the device was not automatically sensed and configured, then run **Add/Remove New Hardware** from the control panel. If it was sensed and there is a problem, you will see a big yellow exclamation point.

Under NT Server, choose:

Start/Settings/Control Panel/Network/Adapters

If the device was not automatically sensed and configured, then run **Add New Hardware** from the control panel. If it was sensed and there is a problem, try removing it, rebooting the computer, and running **Add New Hardware** again.

## REFERENCES

The author's earlier book, *TCP/IP: Architecture, Protocols, and Implementation* (second edition, McGraw-Hill 1996) contains many more details about the TCP/IP family of protocols. For the most up-to-date information, check the RFC documents published at the InterNIC and replicated at other network information centers.

# The Internet Router

This chapter is devoted to configuring the router interface that connects a site to an Internet Service Provider (ISP).

All of the wide area technologies in this chapter are based on "serial" communications. "Serial" simply means that bits are sent out in sequence, rather than in parallel. As illustrated in Figure 9.1, a site's wide area serial interface might be used to:

- Connect directly to an ISP router by means of a dial-up or leased line or Asynchronous Transfer Mode (ATM) service.

- Connect to a frame relay network via a digital dial-up or a leased line. Data would be forwarded across the frame relay network and onto a leased line that connects the ISP to the frame relay network.

- Connect directly to a Switched Multimegabit Data Service (SMDS) network. Data would be forwarded across the SMDS network to an ISP router.

Data leaving a serial interface needs to be formatted differently depending on which of these you will do. Sophisticated routers allow you to configure a serial interface so that your data will be formatted appropriately.

**Figure 9.1**
Methods of
connecting to an ISP.

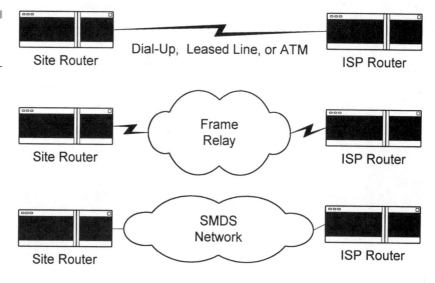

**Figure 9.1**
Methods of
connecting to an ISP.

# Asynchronous versus Synchronous Communication

If you have ever used a dial-up modem, then you have used *asynchronous transmission* across an ordinary phone call. Recall that "Plain Old Telephone Service" sometimes is called "POTS" for short. When you use a modem, data is transmitted one byte at a time. The way that it works is:

- The line is idle until a byte needs to be sent.
- A synchronizing *start bit* is sent that announces that a byte will follow.
- The 8-bit byte is transmitted.
- One or two *stop bits* follow.
- The line returns to idle again.

In contrast, *synchronous transmission* is strictly timed and is based on special digital telephony services. If you have a synchronous connection to the Internet, it will use a timing signal that is emitted by digital telephone system equipment. Every time interval contains a bit. When a system does not have data to send, it will emit a special idle bit pattern.

In this chapter, we will examine parameters that are used for ordinary asynchronous dial-up and for several types of synchronous digital communications.

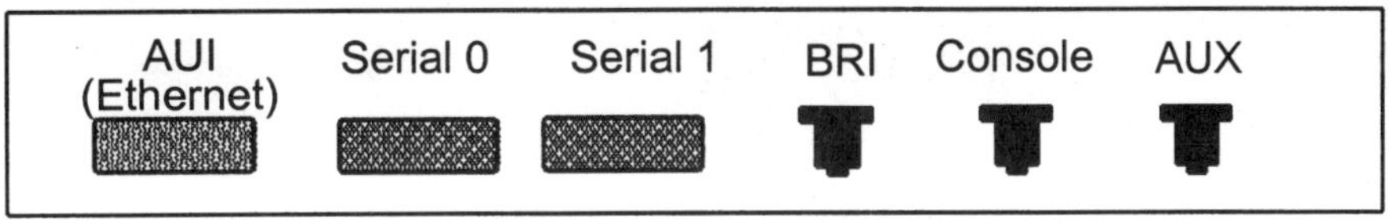

**Figure 9.2**
Sample router ports.

| AUI (Ethernet) | Serial 0 | Serial 1 | BRI | Console | AUX |

## Ports

The back of your personal computer has hardware connectors called *ports* into which you plug your keyboard, mouse, monitor, and printer. It also has communications ports. There usually are two asynchronous serial ports that can be used for dial-up sessions and a LAN port for your Ethernet or Token-Ring connection.

The back of a router has one port for the console terminal or PC that is used to configure and monitor the router. The remaining ports are used for network communications. Figure 9.2 illustrates a router layout that includes one Ethernet port, two synchronous serial ports, a synchronous serial port for ISDN communications (labeled BRI for *Basic Rate Interface* in the figure), an asynchronous port for a console terminal, and an auxiliary asynchronous port. The auxiliary port can be connected to a modem that either is reserved for dial-up by a router administrator, or else could be used for a low speed dial-up connection to or from a remote router.[1]

*Note:* It is extremely unfortunate that the Berkeley developers who integrated TCP/IP into Unix chose to reuse the term "port" to mean something completely different—namely, a numeric identifier for an application session.

## Ports versus Interfaces

Today, even very low-end routers provide synchronous serial ports that can transmit and deliver more than 1.5 million bits per second. You might wish to use one part of this bandwidth for your Internet connection and another part of the bandwidth for a connection to another site.

In the top part of Figure 9.3, a port is completely dedicated to communicating with an ISP. The port corresponds to a single interface. In the lower part of Figure 9.3, the bandwidth for a single port has been split

---

[1]This example is based on the ports available on a Cisco 2503 router.

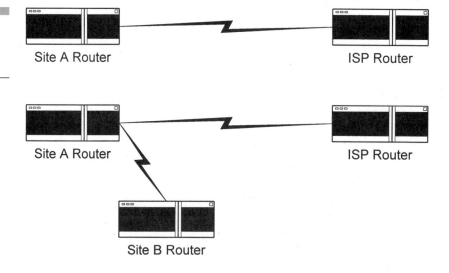

**Figure 9.3**
Using a single port
for one or two
interfaces.

into two parts. Each part is treated as a distinct interface, and is configured separately.

If a port will be used for two or more interfaces, then your first configuration chore for the port will be to define how much bandwidth should be allocated for each interface. Later we will see how this is done.

# Configuration Overview

When you plug a console into your router for the first time, you will need to enter some simple start-up information:

- The name of your router.

- One or more administrative passwords for your router. This will protect the router in the future. No one will be able to access the router to view or change its configuration without authentication.

- Address of a Domain Name Server.

A Domain Name Server is needed because routers occasionally need to perform DNS lookups. For example, routers include software that allows an operator who is connected to the router to start a *telnet* session that connects to another system. DNS lookups would be performed on behalf of a *telnet* user.

Next you will need to configure your local area network interface(s) and the wide area interface that will send and receive Internet traffic. In this chapter, we will examine only the wide area interface.

Some types of parameters are common to all types of wide area interfaces. These include:

- IP address and subnet mask
- Maximum Transmission Unit
- Encapsulation method
- Whether to use dynamic routing
- Routing protocol and metrics
- Transmit queue size
- Traffic filtering

All but the last of these parameters are described in the sections below. The discussion of traffic filtering is deferred until the end of this chapter.

# IP Address and Subnet Mask

According to the rules of TCP/IP, any router interface needs to be assigned an IP address and subnet mask. However, some router vendors allow you to skip IP addresses and masks on point-to-point wide area interfaces. A Service Provider that wants to conserve its pool of addresses might decide to use this feature. However, for testing purposes, it is a good idea to use IP addresses on the link that connects you to the Internet.

# Maximum Transmission Unit (MTU)

The *Maximum Transmission Unit* or MTU is the size of the biggest datagram that may be sent across a link. The default size for wide area links is 1500 bytes, and this is the size that normally is used.[2] A smaller size could lead to serious performance problems.

# Encapsulation Method

In Chapter 8, we explained that data is packaged into frames for data transmission across a link. An encapsulation method is just a frame format and the set of associated rules for transmitting frames. Wide area encapsulation methods include:

- PPP (Point-to-Point Protocol)
- HDLC (High Level Data Link Control)
- Frame Relay
- SMDS (Switched Multimegabit Data Service)
- ATM AAL5 (Asynchronous Transfer Mode Adaptation Layer 5)
- X.25

If your router connects to an ISP router by means of an asynchronous modem or ISDN dial-up, then PPP encapsulation will be used.

For an ordinary leased line connection, either PPP or HDLC encapsulation can be used. PPP is a standard and its use across leased line connections is growing. Each router vendor has developed its own proprietary form of HDLC with some convenient special features. If the same type of router is used at both ends of the link, then the vendor's proprietary HDLC will be used if the ISP prefers to do so.

As you might expect, frame relay or SMDS encapsulation is used if your router connects to an ISP router via a frame relay or SMDS public network.[3]

ATM AAL5 encapsulation is used to connect via the flexible ATM telephony service, which can support a very broad range of bandwidths. ATM service is in its early stages of growth and is not yet widely available.

Prior to the advent of frame relay, X.25 public packet switching services used to be a popular way to obtain wide area circuits that cost less than leased lines. The use of X.25 in the United States is relatively rare today, and so it will not be discussed further in this chapter.

---

[2]This matches the biggest MTU size for Ethernet. It can carry 1460 bytes of data plus a 20-byte TCP header and a 20-byte IP header.

[3]An encapsulation that combines PPP with frame relay has been defined. This is useful for access via a switched frame relay circuit, which is set up on demand like a telephone call.

# Frames and Frames

Earlier, we mentioned how the term "port" has been abused by using the same word to mean:

- A physical connector
- An identifier used to deliver UDP or TCP data to an application

The word "frame" presents a similar problem. As was stated in Chapter 8, in the data communications world, a frame is an organized chunk of data that is sent as a unit across a link. A frame consists of a header, an information payload (e.g., an IP datagram), and a trailing frame check sequence.

As we shall see later in this chapter, the word "frame" also is used in the telephony world, and it means something entirely different. It describes the way that bits are organized to be handled by digital telephone switching equipment.

Both usages will appear in this chapter, but it will be clear from the context which one is relevant. Data communication frames are produced by a host or router interface. Recall that the frame format also is called the encapsulation method. Telephony framing is handled by special equipment—channel service units (CSUs) and data service units (DSUs) that connect routers to leased lines. Usually these items are bundled into one box called a CSU/DSU. Figure 9.4 shows a router connecting to a digital telephone line via a CSU/DSU.

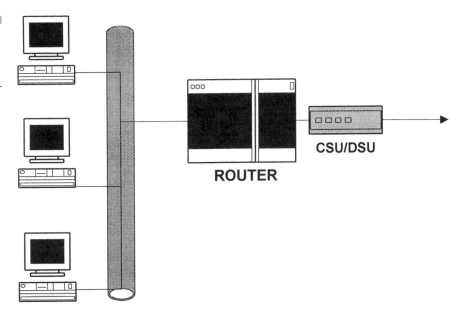

**Figure 9.4**
Connecting to a digital telephone line via a CSU/DSU.

**Figure 9.5**
A simple network
connection to the
Internet.

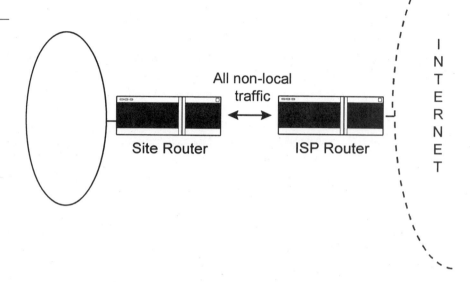

# Whether to Use Dynamic Routing

Routing is dynamic when routers exchange information about the current layout and status of the network. When dynamic routing is used, routers periodically exchange information across your Internet link.

If, as shown in Figure 9.5, you have a simple network and a single connection to an ISP, there is no reason to use dynamic routing. The ISP's router can be manually configured to route to your network, and your router can be configured to route all non-local traffic to the ISP by default.

Even large, complicated sites usually have a very simple connection to the Internet. Often only one LAN is visible to the Internet. The rest of the site is hidden from external view.

However, there are some cases in which a dynamic routing protocol is needed. The choice of protocol is up to your ISP.

## Using RIP-II

Some ISPs wish to use a routing protocol when a single Class C network number has been subnetted among several small sites, each operated by a different customer. This is a convenient way for the ISP to check that correct addressing is being used at each customer site. Only very basic information is needed, and so the ISP may choose to use the simplest protocol, RIP-II. RIP stands for *Routing Information Protocol*.

## More Complex Routing Protocols

Although it is not commonplace, there are some very large sites that have multiple connections to the Internet, sometimes with more than one ISP. Such a site might exchange routing information with its ISP(s) using a sophisticated protocol such as Open Shortest Path First (OSPF), Enhanced IGRP (EIGRP), or even Border Gateway Protocol 4 (BGP4).

If a routing protocol is used, then the interface to the Internet will need to be configured to use this protocol. The interface also will need to be configured with one or more *routing metrics*. A routing metric is a measurement that enables a router to evaluate the efficiency of a particular path. Metrics are used to compute the best route to various parts of the site. Metrics include values such as hop count, bandwidth, and delay.

Most sites will have a single connection to the Internet, and will simply configure a default route that directs all outgoing traffic to the Internet link. Hence we will defer a discussion of routing protocols to Appendix B.

## Transmit Queue Size

Unless you can afford a massive leased line link, your wide area pipeline will have a lot less bandwidth than your LAN (or LANs) may carry. Occasionally, more traffic will be directed to the Internet than can be transmitted immediately. Datagrams will be held in the router's memory, awaiting their turn to be transmitted.

An interface's *transmit queue size* is the maximum number of datagrams that can be held waiting to be sent out of that interface. If the queue fills up, then some datagrams will be discarded. It is a good idea to monitor your router's memory utilization and check the number of datagrams that are discarded for each interface. Making a good allocation of memory can improve performance substantially. For example, interfaces for slow links usually need much more memory than interfaces for fast links.

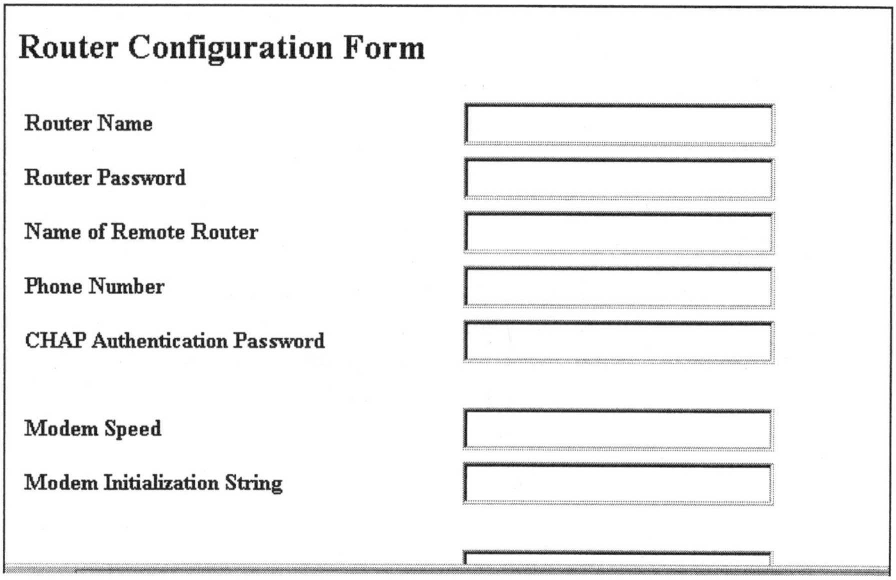

**Figure 9.6**
Using a Web form to
configure a router.

## Router Configuration Form

Router Name

Router Password

Name of Remote Router

Phone Number

CHAP Authentication Password

Modem Speed

Modem Initialization String

# Configuration by Browser

Router configuration used to be an arcane art, requiring an administrator to learn a lot of vendor-specific commands. Many vendors now simplify the configuration by setting a lot of parameters to common default values, and enabling you to enter the essential information on a Web browser data entry form. Figure 9.6 shows what such a form might look like.

The good news is that these forms will do a good job for the most common configurations. However, if you want to change one of the defaulted parameter values, you have to go back to your router command reference.

# Configuring a Dial-Up Interface

Dial-up connections provide low-cost access to the Internet. In Chapter 5 we discussed using Plain Old Telephone Service (POTS) dial-up or the higher-performance digital ISDN dial-up service.

There are several ways to contract for dial-up service for a LAN. You may contract for unlimited service, service during business hours (9–5 Monday through Friday), or some fixed number of hours per month. Parameters needed for POTS and ISDN dial-up include:

- IP address and subnet mask
- Addresses of Domain Name Servers
- Speed (kilobits per second)
- Encapsulation method (PPP)
- PPP password to be used
- Compression of TCP and IP headers
- Enable dynamic routing or not
- Telephone number(s) to call
- Use of dial on demand
- Timeouts

Some of the parameters needed for POTS and ISDN are discussed in the following sections.

# Encapsulation Method

PPP is the standard encapsulation method used for dial-up connections. PPP's support for authentication is an important benefit when using a dial-up link. ISPs like to use the PPP challenge handshake (CHAP) for robust, secure authentication. CHAP will be described in Chapter 10.

# Compression of TCP and IP Headers

Header compression[4] is one of the options that can (and should) be selected for TCP sessions on a dial-up link. TCP and IP headers each are 20 (or more) bytes long. During a TCP session, many fields in the TCP and IP headers change either little or not at all. After the initial messages in a session have

---

[4]Also called Van Jacobson header compression, named after its author.

been sent, subsequent TCP and IP header information is compressed by sending only the changes.

Since the combined 40-byte headers are reduced to 3 to 5 bytes, the saving is substantial, especially when the size of the data payload is small. Extra throughput is gained at the cost of some extra processing that is performed at both ends of the link. For the relatively slow speeds available with POTS or ISDN, this processing is insignificant.

Header compression also can be used on leased serial 56/64-Kbps or T1 lines. This improves performance for the slow 56/64-Kbps link. However, for a T1 link attached to a router that can forward its traffic very quickly, the extra processing might actually degrade performance.

## Use of Dial on Demand

If you have to pay a surcharge to your ISP when you exceed a fixed time limit or if you are using ISDN access that involves telephone company per-minute charges, then you may want to configure your router to perform dial on demand. This means that a call will be set up when a client attempts to connect to an Internet server. If the line is idle, then the call will be terminated.

## Timeouts

You will need to set a timeout on waiting for a connection to complete, a limit to the number of retries, and a waiting period after which the call is tried again. If you are using dial on demand, a timeout needs to be set so that the connection will be dropped when there is an idle period.

## POTS Dial-Up

An asynchronous router port is used for POTS dial-up access. Standard router asynchronous interfaces can operate at up to 115,200 bits per second. (You actually achieve a lower speed because your modem cannot send and receive this fast.) Most routers also have an auxiliary serial port that primarily is intended to be used by administrators who need to access the router by dialing into a modem connected to this port. The auxiliary port speed usually is at most 38,400 bits per second.

Your router will need to be configured with a script that sends a setup string to your modem and guides the dial-up process.

# ISDN

ISDN is a dial-up service that is based on digital technology. It is designed to support high-quality digital transmission at 64 or 128 kilobits per second. You might actually be limited to 56 or 112 kilobits per second. We will explain this later, in the section entitled "Digital Transmission."

If you wish to use ISDN, then you should obtain a router that has a built-in ISDN interface. Physically, the router port looks like an RJ-45 outlet. Instead of a modem, you need a Network Termination type 1 (NT1) box. Your telephone provider probably will supply this and attach it to your wall. You connect the router's ISDN port to the NT1 with an RJ-45 cable.

In the telephony world, configuration is called *provisioning*. In the sections that follow, we will look at some of the information needed to provision an ISDN connection. Recall that an ISDN connection consists of two 64-kilobit-per-second B channels that can be used to transmit data, and a 16-Kbps D channel that is used for call setup and other signaling.

# Type of Switch and Service

In Chapter 5, we mentioned that telephone switch vendors have implemented ISDN digital dial-up many different ways. AT&T built the 5ESS Custom switch. Northern Telecom built the DMS-100. Different rules had to be followed when communicating with each of these switches. Later, Bellcore defined a service standard called National ISDN 1 (NI-1) to promote interworking. Some central offices support NI-1 on AT&T, Northern Telecom, or other switches instead of the proprietary service interfaces.

In Europe, ISDN setup also differs according to the type of switch used. Common European switches include Basic-net3, 1TR6, and VN3.

You need to ask your telephone company what type of switch and ISDN service they support. Make sure that you acquire a router that is compatible. Also check your router vendor's documentation to find out about the ISDN capabilities that you need to request from your telephone company in order to guarantee that the router can be connected to the service. The capabilities that you need to ask for will be different for each switch type.

# Telephone Numbers

Your telephone company will assign one or two telephone numbers for your ISDN service. They will refer to these as Local Directory Numbers or LDNs. One number is assigned for 5ESS Custom service if the line will be used for data only. (A second number is available if voice service also is desired.) Two numbers that can be used for either voice or data are assigned for Northern Telecom DMS-100. Two numbers can be used with the standard NI-1 service.

You will need to decide whether you want to use both channels for data only, or whether you want the option of choosing voice or data.

# Service Profile Identifiers

ISDN can provide a lot of different options and features. For example:

■ Your B channels might be restricted to data only, or might be able to support voice as well.

■ If you accept incoming data calls, you might want to set up an automatic assignment of incoming calls to a free channel. This is called a *hunt group*. For example, if B channel 1 is busy, the phone switch could automatically assign an incoming call to B channel 2.

When you order your service, you may be able to choose from a number of options of this type. The telephone company will assign you one or two *Service Profile Identifiers* (SPIDs) that need to be configured into your router. At start-up time, your router will send the SPID to the central office as part of the initialization process. The SPID will serve as an index that points to the service choices in your contract.

As used today, the SPID is basically a telephone number that optionally has an extra ID tacked onto the end. When you connect a router to an ISDN line, your provider will assign 0, 1, or 2 SPIDs to your ISDN interface. If you have two telephone numbers and your telephone company requires SPIDs, you would configure a separate SPID value for each number.

Unfortunately, many different formats are used for SPIDs. The format depends on the equipment and service setup. You will have to check with your telephone company to get the correct format. Table 9.1 lists SPID formats. It also includes a few examples based on fictitious telephone numbers. The good news is that these SPIDs will be entered once and then you can forget about them.

**TABLE 9.1**

SPID Formats

| Generic National ISDN-1 (NI-1) and ISDN-2 (NI-2) format | |
|---|---|
| aaannnnnnnsstt | aaannnnnnn=area code and 7-digit telephone number<br>ss=sharing terminal identifier, usually 01 for a router<br>tt=terminal ID code, usually 01 for a router<br>Example: 99377755430101 |
| National ISDN-1 (NI-1) service from an AT&T 5ESS switch | |
| 01nnnnnnn0tt | nnnnnnn=7-digit telephone number<br>tt=terminal ID code (Range is 00 to 62. 00 is a good choice.) |
| AT&T Custom Multipoint service | |
| 01nnnnnnn0 | nnnnnnn=7-digit telephone number |
| AT&T Custom Point-to-Point service | |
| | There are no SPIDs for AT&T Custom Point-to-Point service. |
| National ISDN-1 (NI-1) service from a Northern Telecom DMS-100 switch | |
| aaannnnnnnsstt | aaannnnnnn=area code and 7-digit telephone number<br>ss=01 for the first telephone number, 02 for the second<br>tt=ID in the range 00 to 62. 00 should work<br>Example: 99377755430100, 99377755440200 |
| DMS-100 Custom service from a Northern Telecom DMS-100 switch | |
| aaannnnnnnss | aaannnnnnn=area code and 7-digit telephone number<br>ss=01 for the first telephone number, 02 for the second |

# ISDN Parameters

As you can tell from the earlier discussion, ISDN configuration is very product specific. The information that you must enter depends on:

- Your router product
- The type of ISDN switch at the central office
- The kind of service you have ordered

ISDN customers need to have a lot of patience and time, or else use consultant services to get communications running.

The first step in preparing to configure your interface is to write down the information that you need:

- ISDN switch type used by your central office (in North America, typically 5ESS, DMS, or NI-1)

- Your ISDN Service Profile Identifiers (SPIDs)
- Your ISDN local directory numbers (LDNs)

## Digital Telephone Services

When the backbone of the U.S. telephone system was converted to digital, each voice call was transformed into 64,000 bits per second of binary data.[5] Groups of twenty-four calls were bundled together into a T1 carrier line and transmitted between telephone switches that are sprinkled across the continent.

Telephone companies sell leased line services based on connecting customer premises to the digital telephone backbone.

## Using Leased Lines

In the sections that follow, we are going to discuss ways to connect your router to an ISP router that involve the use of leased lines. We will examine:

- Leased lines that connect directly to the remote router.
- Leased lines that connect to a frame relay network.

  Leased lines are available at many different speeds:[6]

- 56, 64, 128, 256, 384, and 512 kilobits per second.
- 1.5, 3, 6, 9, 12, 15, 18, 21, ... 45 megabits per second.

## Why 56 Kbps?

Sometimes when you request 64-kilobit-per-second service, you actually will receive 56 kilobits per second. This is because of the way that some older telephony equipment works.

The United States telephone system adopted the *Alternate Mark Inversion* (AMI) to encode 0s and 1s onto telephone wires. AMI represents 0s as zero

---

[5]This is formally called a Digital Signaling level 0 (DS0) channel.

[6]Different speed levels are used outside of North America and Japan.

voltages and 1s as alternatively positive and negative voltages. For example, 01110101 would be transmitted as:

$$0 + - + 0 - 0 +$$

Unfortunately, long strings of 0s cause transmission equipment to lose timing. Older telephone equipment prevented this from happening by inserting a 1 into every byte of digital data. In other words, you only get to use 7 out of every 8 bits in a byte. This is the reason that in some locations, you get 56 Kbps instead of 64 Kbps.

Up-to-date telephone equipment solves the 0s problem by replacing a sequence of eight 0s with a special 8-bit pattern that cannot be confused with normal data, because it contains a $- -$ or a $+ +$ pattern. This is called *bipolar 8 zero substitution* (B8ZS).

When zero substitution is used across a telephone line, it is called a *clear channel line*. You can transmit data across a clear channel line at 64 Kbps.

## Router Serial Ports and CSU/DSUs

When you want to connect a router to an ordinary POTS telephone line, you need to use a modem. When you want to connect a serial digital port on a router to a 56/64-Kbps or T1 digital telephone line, you need to use a box containing a *Channel Service Unit* and *Data Service Unit* (CSU/DSU). The CSU function in this box provides physical connectivity to the telephone system and supports loopback testing. The DSU matches the local clocking of outgoing bits to the timing of incoming bits coming from the telephone network.

Routers have digital serial ports that are designed to be used with digital telephone lines. The same serial port can be used to interface to a 64,000-bit-per-second line or to a T1 line. The speed will depend on whether a 64-Kbps CSU/DSU or a T1 CSU/DSU is used, because the CSU/DSU unit will provide the bit clocking.

## Configuring a Leased 56/64-Kbps Serial Interface

Configuring this type of serial interface is straightforward. As for any interface, the common parameters described in the earlier section entitled "Configuration Overview" (IP address, subnet mask, etc.) need to be entered.

Some routers have a built-in CSU/DSU and are preconfigured by default so that the interface is all ready to go. Check out your default configuration. You may need to modify:

- Line speed (56 or 64 kilobits per second)
- Clock source (should be the network by default)
- Whether to use scrambled data coding

Scrambled coding is an option that sometimes is used on 64-Kbps lines. The scrambled format prevents the telephone company from generating error control codes.

# Configuring T1 and Fractional T1

Below we list parameters for a T1 interface. T1-specific parameters will be explained in the sections that follow.

- Telephony framing (Superframe or Extended Superframe)
- Call slots to be used
- Speed for each slot
- Telephony line code (AMI or B8ZS)
- Use of signal inversion
- Clock source (the network, except during testing)
- Encapsulation (PPP or proprietary HDLC)
- Enable data inversion

# T1 Terminology and Technology

Recall that a voice call is transmitted digitally as 64,000 bits per second. A telephone company combines bunches of calls together across its wires. Twenty-four calls are sent across a T1 carrier line by transmitting one byte of data for each call in turn. After call number twenty-four, the order wraps around to call one again.

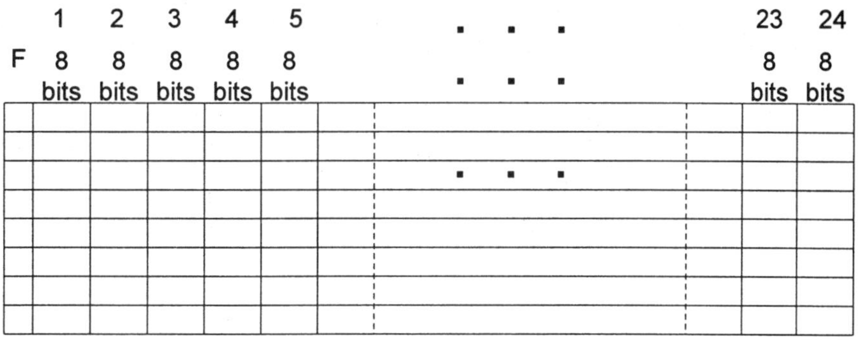

**Figure 9.7**
T1 framing.

| | | 1 | 2 | 3 | 4 | 5 | | . | . | . | | 23 | 24 |
|---|---|---|---|---|---|---|---|---|---|---|---|---|---|
| | F | 8 bits | 8 bits | 8 bits | 8 bits | 8 bits | | . | . | . | | 8 bits | 8 bits |

Figure 9.7 / T1 framing.

# T1 Framing

Bits on a T1 line are sent out according to strict timing. Timing is established by the telephone network. Each of the twenty-four calls is allocated a fixed time slot, as shown in Figure 9.7. Each time slot is called a channel.

It is very important to know where each sequence of call slots begins and ends. To do this, an extra telephony "framing" bit is placed at the head of each row. The telephony framing bits repeat a distinctive pattern that enables the transmission equipment to keep everything in its place. Each of the rows in Figure 9.7 is called a telephony frame.[7]

Originally, the extra framing bits were set in the pattern 100011011100. This pattern was repeated every twelve frames and a set of twelve frames was called a *D4 Superframe*. Most telephone companies have moved to a new pattern that is spread across twenty-four frames. This newer method of framing is called *Extended Superframes* or ESF.

Before configuring a T1 interface, you need to find out whether Superframes or Extended Superframes are used on your telephone line.

A single frame contains 193 bits. Eight thousand frames are transmitted per second, giving a total of 1,544,000 bits per second. If a full T1 link were used as a raw data transmission pipe, then all of this bandwidth could be devoted to data transfer. However, your Internet link will be divided into channels and the framing bits will be used for alignment, so that the maximum throughput is $192 \times 8000 = 1,536,000$ bits per second.

---

[7]Note that these are very different from the data communications frames that we described in Chapter 7. This is another unfortunate case of the same word being used to mean very different things.

# Fractional T1

Service Providers often offer fractional T1 access. This means that they will permit you to use just a portion of your T1 bandwidth to send data and charge according to the portion used. In fractional T1, you configure a group of time slots as an interface that will be actively transmitting data. Bits in the remaining time slots are ignored.

# Configuring T1 Channel Groups

If you have contracted for a full T1 service, then all twenty-four of your T1 channels will be grouped together and treated as a single communications interface.

If you have contracted for fractional T1, then you need to coordinate with your provider and select the group of channels that will be used to carry your data. Table 9.2 shows the number of channels needed for each level of fractional T1 service. Note that once a customer's requirement goes beyond 512 kilobits per second, full T1 is an economical choice, so normally no further intermediate levels are offered.

For example, if you needed a 256-kilobit-per-second connection, then you might create an interface by grouping channels 5 through 8 together. However, any set of channels might be selected: for example, 1, 3, 8, and 10. The channels do not need to be contiguous.

In some cases, you may be restricted to 56 kilobits per second for each call slot instead of receiving the full 64 kilobits per second.

**TABLE 9.2**

Channels for Fractional T1

| Number of Channels | Bandwidth in Kilobits per Second |
|:---:|:---:|
| 2 | 128 |
| 4 | 256 |
| 6 | 384 |
| 8 | 512 |

## Signal Inversion

If you configured the routers at each end of the line to *invert* their T1 signals—transmitting each 1 as a 0, and each 0 as a 1—this would prevent long strings of 0s from occurring. Unfortunately, in most cases, this will not prevent the telephone company from inserting a 1 bit into each byte in the data stream.

## E1

European links use a substitution pattern called High Density Bipolar 3-Zero Maximum Coding (HDB3) to replace a block of four 0s with a pattern containing a bipolar violation.

## Configuring T3 and Fractional T3

An ordinary serial interface cannot support the high data rates provided by T3 or fractional T3 service. A *High-Speed Serial Interface* (HSSI) must be used. A special HSSI cable is needed in order to connect the router interface to a high-speed DSU.[8] The DSU will connect to the T3 transmission line, which is a coaxial cable.

The T3 DSU has to be configured with the high-speed telephony format to use: C-bit or M13. The newer C-bit format (or "application" as it is called in telephony) provides superior error reporting and monitoring functions.

If fractional T3 service will be used, then the specific channels that will be active need to be selected. A T3 line carries 672 64-Kbps channels.

Outside of North America and Japan, E3 carriers are used instead of T3 carriers. E3 uses a different framing format and carries a different number of channels.

---

[8]CSU/DSU units also are available to connect a High Speed Serial Interface to ATM or SMDS networks instead of to T3 telephony service.

# Frame Relay Configuration

Recall that a frame relay "cloud" network provides inexpensive access to an ISP because the links within the frame relay network are shared by many customers. Your router connects to a switch in the frame relay network via a leased line. Your traffic will be sent across this line and then will be switched across the frame relay network to a line that connects to an ISP router.[9]

Configuring frame relay is a combination of serial digital line configuration and additional information that is specific to frame relay. An ordinary serial port on the router is used.

If you connect to the frame relay network by a 56- or 64-Kbps line, then you need to connect the serial port to a 56/64-Kbps CSU/DSU, which will provide the appropriate bit timing and interface to the telephone network. If you contract for a fractional T1 rate or T1 rate, then you need to connect the serial port to a T1 CSU/DSU. For fractional T1, you need to identify the T1 slots that will be used for your interface. Just as before, serial telephony parameters such as the line code and the clock source need to be entered.

Your interface must then be configured to encapsulate your data for frame relay. The information in the frame relay header will be used by the frame relay provider's equipment to switch your traffic to its destination and handle it appropriately. Parameters specific to frame relay include:

- DLCI for the circuit
- Discard Eligibility rules
- Whether to use IETF encapsulation
- LMI type

We'll explain the parameters that are specific to the frame relay technology in the sections that follow.

# The Data Link Connection Identifier

A frame relay interface can support many concurrent circuits. Each is assigned a numeric identifier called a Data Link Connection Identifier or

---

[9]If you were a really big customer, then the frame relay provider might locate a switch right at your premises. In this case, the router could be connected directly to the switch instead of to a leased line. This might be the way that the ISP end of the connection is set up.

**Figure 9.8**
DLCIs at the ends of
a frame relay circuit.

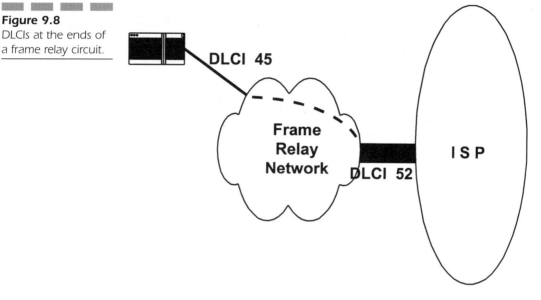

DLCI. Even if you have only one circuit, (i.e., to connect you to the Internet), it will require a DLCI. Every frame leaving your interface has a header that identifies the DLCI of its circuit. Your frame relay service provider will inform you of the DLCI number(s) to be used for your circuit(s).

All of your outgoing traffic will flow down the single line that attaches your site to the frame relay network. When a frame reaches the "cloud," the DLCI is used to route your data onto the correct circuit. The DLCIs at the two ends of a circuit usually are not the same. This is no problem. The frame relay network is aware of the values used at both ends, and will create a frame relay header containing the delivery DLCI before sending it along the destination pipeline, as shown in Figure 9.8.

Since data for several different frame relay circuits can be sent out of a router's serial interface and down the pipeline to a frame relay network, separate circuits are called subinterfaces. Separate parameters are configured for each subinterface. For example, each subinterface has a separate IP address.

# Congestion and Discard Eligibility

If a lot of sites burst data onto the network at the same time, the frame relay network may get congested. The result is that some traffic will be discarded. You can exercise some control over which of your frames are thrown away. The frame header contains a *Discard Eligibility* (DE) flag. Frames with DE = 1 will be discarded in preference to those with DE = 0. You can configure your router with criteria for setting DE = 1.

The frame headers of incoming traffic also contain flags that indicate whether there is congestion on the forward and reverse path to the other end of the circuit. These are called the *Forward Explicit Congestion Notification* (FECN) and *Backward Explicit Congestion Notification* (BECN) flags.

# Encapsulation Method

A simple encapsulation format was defined for frame relay. It consists of a header that contains the DLCI, congestion flags, and discard eligibility flag; the information (IP datagram); and a frame check sequence trailer. No field was included to identify the protocol being carried.

Some users wanted to send traffic for several protocols (e.g., IP, Novell's IPX, and Digital Equipment Corporation's DECnet) across a single circuit. To meet this need, the Internet Engineering Task Force (IETF) defined an additional header field that follows the frame header and identifies the protocol contained in each frame.

The IETF field only is required when multiple protocols are carried. If you are connecting to an ISP, all of your frames will contain IP datagrams, and so the additional header field is not needed. However, you should check with the ISP to make sure that the ISP's router does not expect to see the IETF header.

# The Local Management Interface (LMI)

Your traffic is supposed to be buzzing across the frame relay network, but how can you be sure that everything is working as it should? *Local Management Interface* (LMI) inquiry messages check that the link to the frame relay service is up (via keep-alive messages) and query the status of your virtual circuit(s). LMI messages also can be used to dynamically discover the DLCI that has been assigned to a circuit. There are three different specifications for LMI messages:

■ The Frame Relay Interface joint specification from Northern Telecom, Digital Equipment Corporation, StrataCom, and Cisco Systems. This was created before a standards-based LMI was available.

- The ANSI-adopted Frame Relay specification, T1.617 Annex D.
- The International Telecommunications Union Telecommunication Standardization Sector (ITU-T) Frame Relay specification, Q.933 Annex A.

You will have to check with your frame relay provider to find out which LMI you should use. Many router products support all three.

Along with configuring the type of LMI messages to be used, optionally you also can set LMI timers for sending keep-alives and polling for the status of circuits. However, routers usually have satisfactory default settings for these values.

# Remote IP Address

A frame relay feature called *inverse ARP* allows a frame relay interface to automatically discover the IP address that is being used at the remote end of a circuit, saving the administrator the trouble of tracking the information down and entering it manually. The way that automatic discovery works is that whenever the circuit initializes, your router sends a message to the remote router requesting its IP address.

# Switched Multimegabit Data Service

Switched Multimegabit Data Service (SMDS) is a wide area networking service that provides many levels of throughput. Offered rates usually include gradations up to 155 megabits per second.

Parameters specific to SMDS include:

- Use of SMDS encapsulation
- SMDS address for the interface

# SMDS Encapsulation

The router has the job of packaging data into SMDS frames before transmission. The frames are then passed to an SMDS Data Service Unit that interfaces to the SMDS network. The SMDS DSU combines a *Data Exchange Interface* (DXI) function, which breaks frames into cells, with timing and network interface functions.

## SMDS Addresses

Your SMDS service provider will assign an SMDS address to your interface that plays a role similar to a telephone number. The provider also will assign you an IP address.

The SMDS address consists of 64 bits. The first four bits will be 1100 (hexadecimal C). For the remaining bits, each group of 4 bits represents a decimal digit, for a total of at most 15 decimal digits. Sometimes only 11 digits are used. (When the address is viewed in hexadecimal, the remaining unused bits are filled with 1s.)

Before entering this address, check your router documentation. You may need to enter the number in a dotted format, such as:

C133.4455.1212

C123.4567.1234.3425

## Other SMDS Capabilities

As noted previously, SMDS can be used to set up virtual wide area LANs. Datagrams can be sent between any pair of nodes on the virtual LAN, and data can be multicast from one system to a group of other systems. These functions will not be relevant when SMDS is used for simple connectivity to an ISP router.

## 10-Megabit Service

Although currently available at only certain locations, 10-megabit-per-second wide area accessibility is expanding. There are many ways in which this service may be provided. One method that can be used is to connect to the ISP via an Asynchronous Transfer Mode (ATM) service. Recall that ATM is a telephone service that offers many different bandwidth options, including very high speeds.

Figure 9.9 shows a configuration currently offered by Internet Service Provider UUNET. Frames actually are *bridged* from one site to another. This means that everything between the two routers is treated like a single local area network.

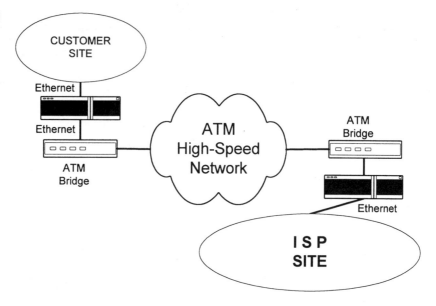

**Figure 9.9**
A bridged connection across an ATM network.

The router at the customer site interfaces to the local site via an ordinary Ethernet connection. Then Ethernet frames are encapsulated in ATM frames and are sent down a link that connects to an ATM service. They are transported across the ATM network to a bridge at the ISP and delivered as Ethernet frames.

Your Service Provider will set up your ATM equipment and interface, but understanding ATM terminology may be helpful. The standard interface used between a router and communications equipment directly connected to an ATM network is a Data Exchange Interface (DXI). This is used in place of the CSU/DSUs that were described earlier.

An ATM connection is associated with a particular telephone system channel, and multiple channels that have the same destination are bundled together into a virtual path. A circuit is identified by two numbers: the Virtual Path Identifier (VPI) and the Virtual Channel Identifier (VCI). These numbers must be entered when configuring the circuit. The other important parameter is the frame encapsulation that will be used. This will be ATM Adaptation Layer 5 (AAL5).

# Traffic Filtering

Filtering is the process of examining traffic entering and leaving your site, and blocking items that may present a security problem. If you connect to the

**TABLE 9.3**

Sample Filtering
Rules

| Direction | Protocol | Source IP | Destination IP | Source Port | Destination Port | Can Open |
|-----------|----------|-----------|----------------|-------------|------------------|----------|
| Out | TCP | proxy/DNS | any | >1023 | 21 | yes |
| In | TCP | any external | proxy/DNS | 21 | >1023 | no |
| In | TCP | any external | proxy/DNS | 20 | >1023 | yes |
| Out | TCP | proxy/DNS | any external | >1023 | 20 | no |
| Out | UDP | proxy/DNS | ISP DNS | >1023 | 53 | |
| In | UDP | ISP DNS | proxy/DNS | 53 | >1023 | |
| Out | TCP | proxy/DNS | any external | >1023 | 80 | yes |
| In | TCP | any | proxy/DNS | 80 | >1023 | no |
| Out | TCP | Mail | any | >1023 | 25 | yes |
| In | TCP | any | Mail | 25 | >1023 | no |
| In | TCP | any | Mail | >1023 | 25 | yes |
| Out | TCP | Mail | any | 25 | >1023 | no |

Internet via a router, it is essential to use one that provides good filtering capabilities. You will need to configure the router with a list of traffic that is allowed to enter and exit each interface, and deny access to all other traffic.

Table 9.3 illustrates a way to tabulate a filtering plan for TCP and UDP traffic. After entering all of your conditions and making sure that they are in the correct order, you can translate them into the commands recognized by your router.

Some sample entries that correspond to the configuration in Figure 9.10 are included. In the figure, internal computers get information from the Internet via a proxy server. The proxy server also acts as a Domain Name Server.

The table shows a few entries for filtering the interface that connects router A to a link leading to the Internet. The table is not complete. The entries include rules that allow the combined proxy/DNS system to:

■ Set up file transfer sessions with Internet hosts.

■ Accept data connections from Internet file transfer servers.

■ Send DNS queries to external servers.

■ Receive DNS responses from external servers.

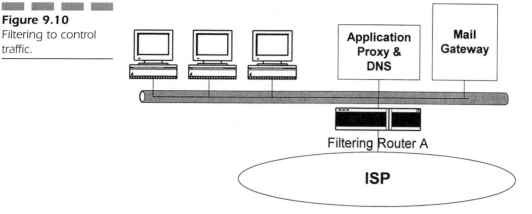

**Figure 9.10**
Filtering to control traffic.

- Set up World Wide Web sessions with Internet servers.
- Forward mail to the Internet.
- Receive mail from the Internet.

The last column indicates whether a request to open a connection from the indicated source to the indicated destination is allowed.[10]

The first four rows require some explanation, When a user wishes to connect to a file transfer server, the user's client opens a *control connection* to port 21 at the file transfer server. This connection is used to carry client commands to the server, and server responses back to the client. The first two rows of the table enable the proxy to initiate control connection with the server and exchange commands and responses.

When a user wants to copy a file from the file transfer server, the normal procedure is that the file transfer server initiates a new connection that is used to transfer the file. Rows 3 and 4 contain the rules that enable a file transfer server to open a data transfer connection to the proxy client. Unfortunately, these rows will permit any external computer to open a connection to the proxy client, as long as the source port is 20 and the destination port is > 1023. (A smart filtering firewall could make sure that the incoming connection really corresponded to a prior client request.)

In the table, we indicated that all outgoing DNS queries will be sent to a DNS server operated by our Internet Service Provider. The local DNS server can be configured to forward all of its queries there. This can be done to cut down on the number of external systems allowed to send traffic through the firewall.

---

[10]The filtering firewall can recognize a request for a new connection because that is the only message for which the TCP ACK flag is 0.

This table allows the proxy system to connect to external Wide World Web servers and to receive their responses. It permits the Mail Gateway to connect to external computers to forward mail out of the site, and allows Internet systems to connect to the Mail Gateway in order to deliver mail to the site.

Examining the entries should illustrate why a smart screening firewall does a much better job. A smart firewall can remember exactly which external host should be allowed to send data to a local host and which should be allowed to open file transfer data connections. It knows which current session each incoming datagram belongs to.

We will take a closer look at security issues in Chapter 10.

# 10

# Security Basics

In spite of sensational news stories about Internet break-ins, today most Internet sites still have not implemented adequate security protection. Internet hackers will try to steal company secrets, corrupt your data, clog the line to your site with junk data, or plant embarrassing pages at your Web site.

The first step in securing your site is to understand the underlying reasons for Internet security problems and to become aware of the solutions that are available.

Internet applications and TCP/IP evolved in an academic research community in which there was an atmosphere of trust. It was like a small town in which people don't bother to lock their doors. It was more important to create applications that furthered communication between people than to check the application software for flaws or security holes.

Today there are technologies that can seal up the holes in Internet security and enable you to perform serious business transactions on the Internet. First, let's take a quick look at the root causes of some of the problems.

# Sniffing

The LAN technologies that are so prevalent today—Ethernet, Token-Ring, and FDDI—were designed to work in a trusted workgroup environment. Data traversing a LAN actually is accessible to any system that is connected to the LAN. PC or workstation users can configure their own interface cards to eavesdrop on traffic as it passes. This practice is called "sniffing." Sniffing can be prevented by using switched rather than broadcast LAN implementations. Switching traffic to every single desktop is still somewhat expensive, so that broadcast-style links remain prevalent.

Your Internet traffic may pass through several intermediate networks, and you cannot assume that all of these will be safe. There have been several notorious sniffing attacks on the Internet. You cannot assume that the contents of email messages or other transmitted data will remain confidential.

# Opening Up Access

Traveling personnel will want to pick up their electronic mail from their home electronic mail server, or access data from their internal network. It is tempting to open up access to internal systems via the Internet because of the convenience and low cost.

But usernames, passwords, and corporate data may be picked up via Internet sniffers. Even without sniffing, hackers can mount a password attack against your host systems, automatically trying out millions of passwords until one of them works.

# System Insecurity

Since your site will probably expose at least a mail server and an application proxy to the Internet, keep in mind that any exposed system can be subjected to Internet attacks.

Vendors tend to ship their operating systems preconfigured to run every type of Internet service. Often vendors include free software that is not robust enough to stand up to attack. New computers also may have some preconfigured guest login accounts set up that have publicly known passwords. Hackers know all of the ways of getting into soft systems that have not been properly prepared for life on the Internet.

# Application Insecurity

Internet applications such as electronic mail and file transfer are big and complicated and were not designed to withstand hacker attacks. On the contrary, anyone can walk up to one of these applications, talk to it, and look for a weakness. For example, in the display below, we use a simple Windows 95 terminal emulation program (*telnet*) to connect directly to port 25 at a remote mail host and talk directly to its electronic mail program.

```
220 mail.abc-corp.net ESMTP Sendmail 8.7.6/8.7.3;
  Sun, 26 Jan 1997 10:23: 04 -0500 (EST)
HELO hackhost.xyz.com
250 mail.abc-corp.net Hello hackhost.xyz.net
[130.192.66.101], pleased to meet you
```

It is important to use the most recent version of an Internet server program because it will include fixes for the known security problems.

# Forging IP Source Addresses (Spoofing)

Keep in mind that the owner of a personal computer can configure it any way that they like, and can run any software they wish. Hackers have sneaked into many sites by configuring their computers with bogus IP addresses. This process is called *spoofing*. Trust should never be based on source IP address alone.

Normally, if a hacker includes a fake IP address in a datagram, the hacker never will see responses because the hacker is not really at that address. However, IP has a feature called source routing, which allows a sender to enclose the path to be followed to and from the destination. As shown in Figure 10.1, responses to source-routed traffic will end up at the hacker's computer.

One of the advantages of the Internet is its flexibility and the way that data paths can change dynamically. A disadvantage from the point of view of security is that unlike a telephone call, which follows a fixed path that is established at call setup time, a datagram's past history cannot be traced, giving hackers the anonymity that helps them to escape detection.

**Figure 10.1**
Spoofing a fake
source IP address
using source routing.

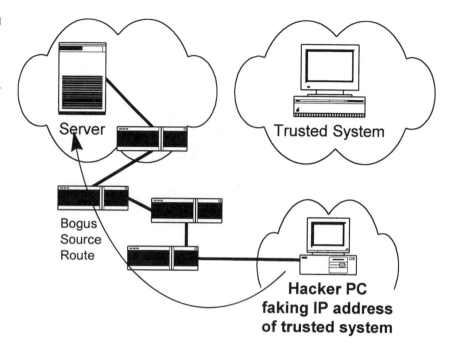

## Forging Electronic Mail

It takes about five minutes to learn how to forge electronic mail. This trick has been used to impersonate a system administrator and send mail that convinces users to change their passwords to specified values. It also was used to launch an attack on the White House mailboxes *president@whitehouse.gov* and *vice-president@whitehouse.gov* that blocked the communications line leading to the mail host system.

Here is the way that it worked. There are thousands of special interest mailing lists that anyone can join. For example, you can subscribe to a chocolate lovers list by sending mail to *listserv@idma.com* and putting "subscribe chocolate" on the first line of the message.

Every mail item sent to the list is forwarded to all of its subscribers. A hacker forged email from the president and vice-president subscribing to thousands of these lists. The result was a glut of traffic heading for the White House. An attack that causes mail to pour onto a site is called *mail bombing*. The effect of this attack was amplified, because the White House system automatically sends out replies, which in this case were delivered to all of the mailing lists, and then were sent back!

Other forms of impersonation have been more direct. Hackers have called network operations centers late at night, claimed to be the president of the company away on a business trip, and have convinced operators to open the company firewall and even provide a new password because the president had "forgotten" their own.

# Denial of Service

A denial of service attack is intended to disrupt access to your site or even activities on your internal network. The email impersonation described above resulted in a denial of service. Denial of service attacks can be simple and blunt. An attacker uses a fake source IP address in order to avoid detection, and writes a program that sends a flood of email items or an avalanche of raw datagrams down the link to the victim's site.

# User Carelessness

The final security exposure is in the hands of your own users. They need to choose good passwords and guard their password information. They need to screen the data that they bring into your site for viruses. They have to understand what needs to be protected, what the rules are, and what legitimate solutions are supported.

Users sometimes open their sites to external communications. People leaving work for the day have been known to set up a live modem at their desktop and configure the desktop to route traffic. Their intention may have been to put in some overtime from home, but they have opened a big back door into your site.

# Solutions

Safeguarding your Internet connection is carried out by:

- Analyzing your requirements.
- Creating a site security plan.
- Creating a site disaster plan.
- Creating a Site Security Handbook and training end users.

- Obtaining firewall hardware and software.
- Implementing new authentication, authorization, and encryption techniques.

Organizations and ISPs also can take some steps to stop attacks that originate within their networks. Outgoing traffic can be scanned to see whether its source address makes sense, and streams of messages intended to damage a recipient can be cut off at the source.

# Requirements

What are you going to do with your connection? Security requirements for simple electronic mail transfer differ from those needed when you will operate your own high-volume Web storefront that promotes online ordering.

Your requirements list should extend beyond the present to cover needs over the next couple of years.

# Creating a Policy Statement

What users want to do may be in conflict with your existing corporate policies on information control and security. These issues need to be resolved. Using an Internet connection is a business choice, not a technical choice. Policies need to be guided by business decision makers who take overall risks and benefits into consideration.

To expedite the process, it is easiest to start by trying to set broad general goals and then work down to the details of exactly what information needs to be protected. The IETF Site Security Handbook can be helpful in this process. It is available at:

*ftp://ds.internic.net/rfc/rfc1244.txt.*

There also is an excellent online paper by Gary Kessler that contains many useful references at:

*http://www.hill.com/library/secure.html*

The end result of the information-gathering stage should be a written policy statement. It is important to publish this statement to all employees. Security restrictions will probably cause some inconvenience to your own users, and without their active cooperation, any security plan can be subverted.

A large organization should bring in its legal department and factor legal exposures into the policy statement. For example, an organization that fails to protect confidential business information can be sued by its stockholders for a breach of fiduciary duty, claiming that there was negligence. If you have databases that might become exposed to the Internet, you will need to set up procedures for careful logging, preservation of original log data disks, and audited authentication of disks. These logs will be an essential part of your proof that appropriate practices were in place. This type of evidence also is needed when you want to prosecute a hacker.

# Creating a Security Requirements Specification

Your security requirements specification translates the policy statement into concrete terms. For example:

- Access to information in databases A, B, and C must be restricted to authorized personnel, and the integrity of the data must be provable.
- The availability and reliability of system X, Y, and Z must not be damaged.

# Creating an Implementation Plan

By this time, you know what needs to be done and are ready to decide how to do it. Security mechanisms include:

- Router filtering
- Firewall systems
- Authentication tools
- Authorization tools
- Encryption tools
- Auditing tools

One of the chores that you need to do is to create a handbook that describes all of the steps that should be taken when your site is under attack or has been damaged. This will include:

- Criteria for disabling servers or shutting down the Internet connection.
- Telephone numbers of ISP security personnel.

■ Telephone numbers of appropriate local staff. This might include legal or public relations personnel as well as technicians.

■ Telephone number for the Computer Emergency Response Team (CERT), which is described later in this chapter.

■ Contact information for other sites that might be affected. For example, your site might have been used as a launch point for an attack against another site. Be ready to consult the InterNIC *whois* database.

Even when you use the best defensive tools that are available, you also need to prepare for the worst. It is important to plan ahead and look at the consequences that would follow if a server became disabled. Frequent backups and the ability to switch to another server quickly need to be a part of your operations plan.

Unfortunately, there is no magic box that will automatically ward off all threats to your site. Security hardware and software products are tools that must be in the hands of a team of knowledgeable security specialists. A large site will need a staff that can monitor your connection to the Internet, have the expertise to detect security problems, and cope with them. This can be costly. In Chapter 6, we discussed the option of outsourcing this job to your ISP.

# Perimeter LAN

In Chapter 5, we introduced a popular configuration called a perimeter LAN that is used at some large sites. Figure 10.2 shows another example of how a perimeter LAN might be set up. Keep in mind that this is an example, not a blueprint. A slightly different arrangement was shown in Chapter 5.

Systems on the perimeter LAN in Figure 10.2 have real Internet addresses that are listed in the public Domain Name System. This LAN is set up as a buffer zone between your site and the Internet. Access rules are enforced by the two filtering routers that guard the boundaries of the perimeter LAN. The router connecting to the Internet is a smart filtering firewall router. (We'll describe its capabilities in a later section.) For extra security, some sites also install filtering software on each of their server hosts.

The perimeter LAN contains the minimum number of systems that are needed to provide the desired services. There are no end-user desktop computers on the perimeter LAN.

Internal users access Internet applications by setting up connections to the proxy firewall server. To reduce the number of DNS queries, you can run a *caching only Domain Name Server* at the proxy host. This means that the

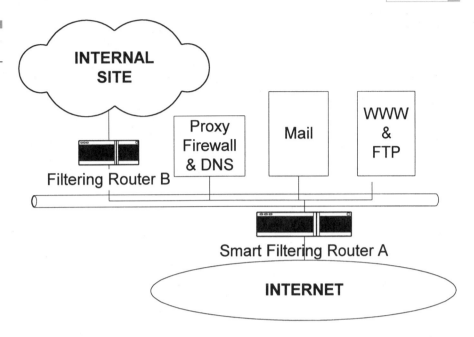

**Figure 10.2**
A perimeter LAN.

server will make queries to the outside world on behalf of your users and will save (cache) the responses on its hard disk for a while. It will not answer any queries about your systems.

External users can access the computer that offers World Wide Web and file transfer services. Mail is forwarded to and from the Internet via the mail server.

# Router Filtering

In Chapter 9, we looked at an example of some filtering rules that might be used to control the flow of information entering and leaving your site. We will take another look at filtering in this section. A filtering router examines datagrams and can reject traffic based on:

- Protocol, such as UDP, ICMP, or a routing protocol
- Source or destination IP address
- Source or destination UDP port
- Source or destination TCP port
- Attempt to open a new TCP connection
- Type of ICMP message

There are a few other criteria, which will be described later. The safest way to filter is to enumerate what you want to allow and forbid all other traffic.

# Filter Checkup

Earlier we mentioned the use of source routes by hackers to fool routers that have been set up to allow some traffic from "safe" locations to enter the site. Prudent managers configure their routers to block all source-routed datagrams. In any case, source IP addresses alone are not a good basis for trust.

For example, it is important to prevent a datagram from sneaking through your router by pretending to come from a source address that is inside your network. You might think that this would automatically be blocked. In fact, when a hacker based a set of attacks on using a forged internal address, it was discovered that some routers could not even be configured to stop this traffic. (Vendors updated their software pretty quickly.)

Some applications and services are known to have the potential for causing or contributing to security problems. It is a good idea to run tests against your filtering routers to check that it blocks attempts to access the items in Table 10.1. The first item, the "little services," have been used in an inventive manner by crackers. When sent any message, the time-of-day, quote-of-the-day, and character generator service each will return an appropriate response to the sender. The echo service just returns exactly what it received.

To understand how hackers caused trouble using these little services, look at Figure 10.2 and imagine what would happen if a hacker managed to get a message to the time-of-day service at the Web/File Transfer host that appeared to come from the echo service at the mail gateway host. The two systems would ping-pong messages back and forth forever. If a hacker set off many interactions of this type, it could destroy all performance on the perimeter LAN.

Note that Table 10.1 recommends blocking some ICMP messages. Some administrators prefer to block out all ICMP messages, since a hacker could try to disrupt a site with a flood of junk ICMP traffic. Recently even the innocent *ping* message was turned into a lethal weapon when it was discovered that oversized *ping* payloads actually could crash many hosts.

You may also wish to screen out all fragmented datagrams. TCP port numbers are very important in making filtering decisions, but the TCP header will be enclosed in only the first fragment. A hacker might try to disrupt your network by streaming in fragments.

**TABLE 10.1** Traffic that Needs to be Excluded.

| Application | Protocol/Port | Description |
| --- | --- | --- |
| Little services | TCP&UDP /7,9,13,17,19 | Echo, time-of-day, quote-of-the-day, character generator, discard. |
| DNS zone transfer | TCP/53 | Prevent outsiders from accessing internal DNS databases to find out about your internal hosts. |
| Trivial file transfer | UDP/69 | Opens up unauthenticated access to files. |
| Finger | TCP&UDP/79 | Provides system and user information. |
| Terminal link | TCP&UDP/87 | A vulnerable private terminal link. |
| Remote Procedure Call Portmapper or RPCbind | TCP&UDP /111 | Identifies the location of certain applications, and also relays queries and responses for some applications. |
| Network File System | TCP&UDP/2049 | File server. |
| Berkeley "r" commands | TCP/512, 513, 514 | Remote execution, remote login, remote command. The latter two were designed to enable access from selected hosts without providing a password. |
| Who | UDP/513 | Shows who is logged in to a remote machine. |
| Print server (lpd) | TCP/515 | Allows a remote user to put files onto the print queue. |
| UUCP service | TCP&UDP /540, 541, 117 | Enables a remote user to copy files or submit commands. |
| Openwindows | TCP&UDP/2000 | Remote application controls the contents of the user interface at a user's system. |
| X windows | TCP&UDP/6000+ | Remote application controls the contents of the user interface at a user's system. |
| NETBIOS services | TCP&UDP /137, 138, 139 | Microsoft networking services could open up access to your files. |
| ICMP redirect messages | ICMP | Spurious messages could confuse your systems about how they should route their traffic. |
| ICMP destination unreachable messages | ICMP | Spurious messages could disrupt existing connections. |
| Unknown. Datagram is fragmented. | TCP&UDP /any port | Spurious messages could clog the network and fill memory at destination systems. |

# Smart Filtering Firewalls

In the past, the filtering provided by most routers was fairly crude. A router had to decide whether to accept or reject a datagram based on the contents of that one datagram. It could not make judgments based on what a user was trying to do. For example, every time that a user copies a file from a file transfer server, a fresh connection normally is opened from port 20 at the server to the client computer. An ordinary filtering router either has to let in all connections that come from port 20 at an outside computer, or block all of them.

For this reason, some sites have replaced or supplemented their routers with full-fledged smart filtering firewall routers (also called smart screening firewalls), as shown in Figure 10.3. A smart screening firewall does not take over the full operation of an application the way that a proxy does, but it keeps track of what goes on in each user's application. For example, the screening firewall would remember that the file transfer user had asked the server to open a connection for a fresh file copy.

Smart screening firewalls are fast and can handle a lot of traffic. They can examine data for a wide range of applications and services, discard according to sophisticated filtering rules, and log suspicious events. Many other features have been added to the best products. They can encrypt information for spec-

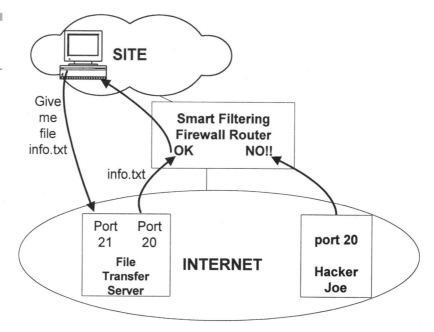

**Figure 10.3**
Difficult decisions for a router.

ified destinations. If you have a big Web site, and want to set up several Web servers, they can distribute incoming Web requests to the various servers, distributing the load.

When a smart screening firewall is used, traffic will be examined carefully, but internal users participate in sessions that are directly connected with Internet computers. Some screening firewalls support Network Address Translation (NAT), which translates internal addresses to addresses taken from a pool of public addresses. You can add another layer to your protective armor by installing filtering and encryption software on each of your servers.

# Application Proxies

Application proxy firewalls were introduced in Chapter 5. A proxy prevents Internet hosts from communicating directly with your internal systems. A proxy participates in the Internet activities of internal users and controls Internet access rights.

If you wish, you can use private enterprise IP addresses across your entire site. (These addresses were described in Chapter 7.) The only system that needs a public address is your application proxy, and this can be assigned by your ISP.

An application proxy provides very robust security protection. It includes fully functioning clients for applications such as the World Wide Web and file transfer, and performs all application activities on behalf of the user. Some proxy systems are bundled with Web, file transfer, and mail gateway services, so that all external functions are performed on a single exposed system.

In Chapter 7, we saw how a browser client can be configured to use a proxy. After the client has been configured, the user can forget about it and use the browser in the normal manner.

Why do some sites prefer screening firewalls to application proxies? Since a proxy actually must execute the application for the user, users are limited to applications supported by the proxy product. An application proxy needs more system resources than a screening firewall, so a screening firewall can handle more traffic if you are fortunate to have a very high-bandwidth link to the Internet.

On the other hand, proxy vendors have been moving quickly to add popular applications to their systems. The debate between proponents of proxies and smart screening firewalls probably will not last very long. Firewall products are evolving quickly, and many of today's products offer users any mixture of proxying and screening that they wish to have.

# Circuit Proxies

A site that wishes to use an application that is not yet implemented by an application proxy, but still wants to avoid direct sessions between internal and external users can use a simple circuit proxy. A circuit proxy provides a very primitive way to separate internal users from the Internet. The user's application opens a connection to the circuit proxy, provides authentication information, and identifies the target host and port. The circuit proxy checks whether the user is authorized to attach to the requested host and port. If so, the circuit proxy opens a session to the target system and then just dutifully copies bytes between the two sessions.

Unlike an application proxy, the circuit proxy does not actually act as a client program. It has no knowledge of what is going on. It just blindly copies data from one session to the other session. But a circuit proxy does hide the identity of the internal system from the Internet.

A free circuit proxy package called SOCKS has been available for several years. Detailed specifications for SOCKS have been proposed as an Internet standard. Free software and more information are available at *http://www.socks.nec.com/*. Some firewall vendors incorporate an implementation of a SOCKS circuit proxy into their products.

# Bastion Hosts

The hosts on your perimeter LAN are in an exposed position and need special care. You should not count on a filtering router or a firewall to keep all of the threats away. Humans who configure computers make errors, and hackers find new ways to fool them. It is wise to apply preventive action to exposed hosts even if the router is supposed to block access. No system should be exposed to the Internet until all operating system and application security leaks have been plugged and it has been subjected to a careful security audit.

We mentioned earlier that hosts usually are shipped ready to run all kinds of dangerous—and usually unnecessary—applications. Your perimeter systems need to be turned into "Bastion" hosts that have been prepared for the rigors of Internet access. This requires:

- Applying the most up-to-date patches to the operating system
- Removing all unnecessary applications
- Installing secure versions of all supported applications

■ Removing or disabling all unnecessary utilities

■ Configuring audits of system and file access, and logging events

In particular, all of the applications and utilities mentioned in Table 10.1 should be removed from Bastion hosts. An NT Server should be set up with an "NTFS" file system, which supports directory and file security. An NT server should not run Server Message Block (SMB) Server, Windows Internet Name Service (WINS) client, or NETBEUI network protocols. As was mentioned earlier, always be prepared for the worst. Back up data often, and have a standby system ready to go if availability counts.

Some vendors sell systems that have been set up as secure bastion hosts. If you are willing to pay extra, you can get an operating system that runs at a higher level of security than ordinary commercial systems and has many extra safeguards built in.

Next we will turn to a set of encryption technologies that are making safe communication a reality, and are being used as the basis of secure business transactions on the Internet.

# Message Digests

A message digest is a special computation on the bits in a message that has some special features:

■ If even one bit of a message is changed, its message digest changes a lot.

■ It is virtually impossible to work backwards from a message digest answer and construct a message that yields that answer.

Sometimes the calculation of a message digest is referred to as a *cryptographic checksum* or a *secure hash*. The MD5 (Message Digest 5) calculation often is used for this purpose. MD5 produces a 128-bit "hash" result. The likelihood that two messages have the same MD5 message digest value is:

1/1,000,000,000,000,000,000,000,000,000,000,000,000,000

Another algorithm called the Secure Hash Algorithm (SHA) was designed by the National Security Agency (NSA) and has been accepted as a national standard. SHA produces a 160-bit hash result.

Message digests can be used to check the integrity of disk files:

■ A message digest is computed for each executable file on a disk, and the answers are stored offline (on a floppy disk or a tape).

■ Periodic recalculations will reveal whether programs or data have been changed.

This is especially useful in order to detect whether someone has planted a *Trojan Horse* program on a disk. A Trojan Horse program is one that performs useful functions, but contains some hidden software that is used to damage or spy on a system. A favorite vandal trick is to replace an ordinary, trusted program with a Trojan Horse version written by the vandal. Message digest calculations ferret these out.

# Combining Message Digests with Secrets

In the sections that follow, we will see how message digests can be used to protect data integrity and to design safe login procedures. This is done by concatenating a secret key[1] with a message prior to computing its digest. Two popular ways of doing that are shown in Figure 10.4.

*If a secret key has been concatenated with data before performing the message digest calculation, someone performing the calculation again*

---

[1]A password or long pass phrase can be used as a key.

**Figure 10.4**
Including a secret in a message digest computation.

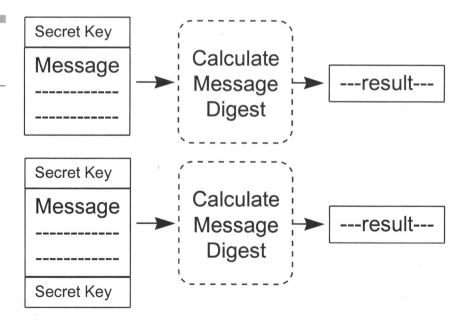

*at a later time cannot get the same answer unless they know the secret key.*

Later, we'll see that message digests also can be used to create reliable electronic signatures.

# Message Digests and Data Integrity

A message digest on data concatenated with a secret key can be used to assure that the information that is sent in an electronic mail message or in some other data transmission has not been altered by a vandal. For convenience, we'll call a digest computed on data concatenated with a secret key a secret digest.

A sender and recipient need to agree on a secret key to be used to create secret digests for the information that they exchange. A sender cannot enclose a correct secret digest with a message unless the secret key is known.

# Authentication

The traditional way that a computer authenticates a user is by asking for a username and password. Sending this information across the Internet is an invitation to a break-in. Encrypting a password before sending it does not help. Anyone sniffing a link along the way can copy the encrypted password and send it whenever they want to.

Far better methods of authentication are available today. The most popular are based on message digests. We'll look at a simple example in the next section.

# Challenge Handshakes

The challenge handshake authentication scheme is based on secret digests. A user's password never is sent across a network. The password (or pass phrase, since longer strings are preferred) is used as the secret key in a digest calculation. As shown in Figure 10.5, the series of steps is:

1. The user sends a username to a host.
2. The host sends a random message to the user. (This message usually includes a timestamp as well as some random data.)

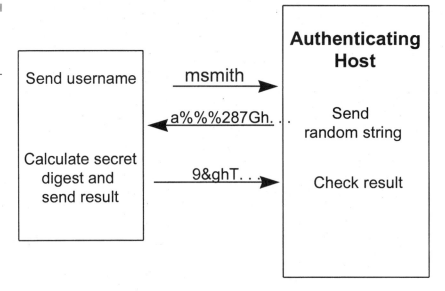

**Figure 10.5**
Authenticating a
client via a challenge
handshake.

3. Both calculate a secret digest based on the message and the user's pass phrase.

4. The user's computer sends the answer to the host.

5. The host compares answers. If the user's computer sent the right answer, then the user is authenticated.

Internet Service Providers use this method to authenticate their dial-up users. The Challenge Handshake Protocol (called *CHAP* for short) is a feature of the Point-to-Point (PPP) protocol used to initiate communication with the ISP's router and carry data to and from the router. Challenge handshakes also have been adopted by businesses that want to provide dial-up access to their users.

# Other Authentication Technologies

Some sites with a need for extra strong authentication use "smart cards" that generate a new, unpredictable access code every minute. The values on a user's card match values that are generated at a security server. A user must enter a Personal Identification Number (PIN) and then will enter the current code displayed on the card.

Other smart cards contain certified identification codes. We'll describe certification a little later. Card vendors are creating smart cards that contain authentication information and encryption processing capabilities, and computer vendors are building keyboards and other plug-in peripherals with insertion slots for these smart cards. A user inserts a card, enters a PIN number, and then can use the card for both authentication and encryption.

Even stronger authentication is based on *biometrics*. A biometric is a measurable physical characteristic that verifies the identity of a person through an automated procedure. Biometrics include characteristics such as retinal scans, fingerprints, voice prints, and hand geometry. This is currently an active development area, and a number of products are on the market.

## Authorization

Authentication deals with validating the identity of a user. Authorization is the process of controlling access to resources depending on this identity.

A growing number of organizations are providing sensitive information via Internet servers. We saw some examples in Chapter 2. The Aetna site enables users to view the current value of their retirement accounts. Security First Network Bank customers can check up on their account balances and transfer funds. Other sites provide information services that are available only to paid subscribers.

Once a server moves from offering free information that is available to any visitor to offering information that varies by user, the server needs robust authorization software. This goes hand in hand with the use of solid operating system platforms that are capable of enforcing authorization rules. Some vendors offer authorization servers, which provide centralized access control. Applications need to be modified in order to cooperate with these servers.

## Privacy and Encryption

The only way to ensure data privacy on the Internet is to encrypt the data. Traditionally, before you could communicate safely with someone, the two of you had to get together and agree on a secret key that would be used to encrypt the data. As shown in Figure 10.6, the same key would be used to decrypt the data. This is called *symmetric* encryption.

Symmetric encryption has been around for a long time, but it has always been difficult to administer. To be really safe, you need a different key for each partner with whom you communicate. Keys need to be changed fairly often. Managing keys can be a difficult problem, especially with the large number of potential partners with whom you might communicate across the Internet.

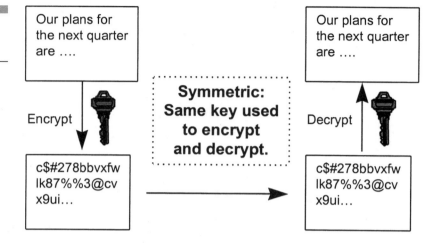

**Figure 10.6**
Symmetric
encryption.

## Secure Tunnels

Site-to-site access via the Internet is hard to beat for convenience and price. Router and firewall vendors have overcome the security problems inherent in Internet communications. They enable you to "tunnel" traffic across the Internet by encrypting data between routers or firewalls, as shown in Figure 10.7.

Even individual users can access their business sites safely across the Internet when their systems have appropriate security software. For example, Microsoft provides free Point-to-Point Tunneling Protocol (PPTP) software for its Windows NT and Windows 95 systems. (However, special router software must be installed in order to use this.) The Point-to-Point Protocol CHAP process is used to authenticate a PPTP user and also is the basis for generating an encryption key for the session. All traffic to and from the site is then encrypted.

There is a proposed IP security standard for authentication and optional encryption of IP traffic between routers, host-to-host, or at the user-to-host level. Products that support IP security are available today, and can be used by sites that need a high level of security.[2]

## Public Keys

A new asymmetric encryption technology was introduced a few years ago. To understand what it does, let's take a look at Figure 10.8. Suppose

---

[2]Sun Microsystems has introduced Simple Key-management for Internet Protocols (SKIP), which facilitates the use of IP security.

**Figure 10.7**
Data passing
through a secure
tunnel.

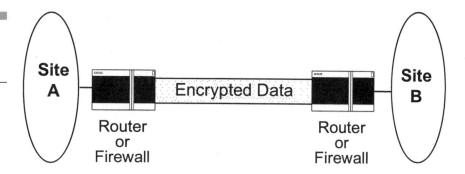

**Figure 10.7**
Data passing
through a secure
tunnel.

**Figure 10.8**
Locking and
unlocking with
separate keys.

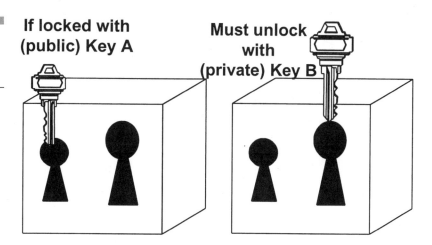

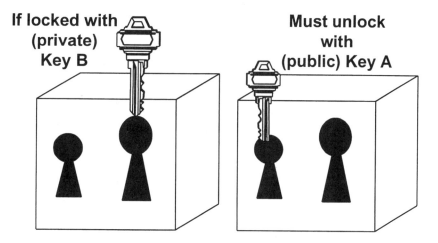

that I have a shatter-proof box that has two different keys, key A and key B.

- If I lock the box with key A, then I must unlock it with key B.
- If I lock the box with key B, then I must unlock it with key A.

Now suppose that I make lots of these boxes and plenty of copies of key A and send them to my business associates. If someone puts something into the box, locks it up with key A, and sends it to me, then no one else will be able to peek into that box. I have the only copy of key B and *only I can unlock these boxes.*

Public key encryption works this way. There is software that you can use to generate a public and private key pair for your own use. You then can distribute your public key to the world. *If someone sends you a message encrypted with your public key, you are the only one who can read it.*

## Combined Encryption

Actually, entire messages are not directly encrypted using a public key. For reasons of efficiency (using a public key involves a lot of computing) and safety (having a lot of encrypted text to work with might help a cracker) a two-step scheme is used. *A fresh random key is generated for each message.* As shown in Figure 10.9, message contents are encrypted using a random key. The random key then is encrypted using the *recipient's* public key and sent along with the message. This is like putting the key into a secure digital

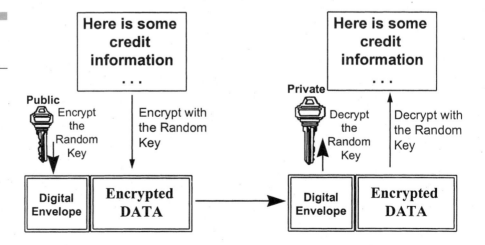

**Figure 10.9**
Combined
encryption.

envelope. The recipient unlocks the digital envelope (using the matching private key), removes the random key, and then decrypts the message.

# Electronic Signatures

Message digests have been combined with public key encryption in a clever scheme that creates reliable *electronic signatures*. I generate my electronic signature by sending you something unique to this message that *only I* can create. That something is the message digest encrypted with my *private* key. You will use my public key to unlock the signature block. A correct encrypted message digest value could only have been created by me—the owner of the matching private key.

This signature will be valid even if the rest of the message is sent in the clear. If anyone changes the message or tampers with the signature, this will be detected when the receiver unlocks the digest and rechecks its calculation.

If total confidentiality is needed, then combined encryption will be used on the message. The electronic signature will be encrypted along with the message contents using the random key that was chosen for this purpose. Your private key will be needed to start the whole message-unlocking process.

# Certification

There is still one element that is missing in order to establish trust in a distributed environment. If I know you, then I can write down my public key and give it to you.

But you will wish to communicate reliably with people whom you have never met and anyone can generate a public/private key pair. How can you be sure of the identity of the owner of a public key that is listed in some public directory? Validating the ownership of a particular public key is at the heart of making Internet authentication reliable. It also is essential for conducting safe commercial transactions.

The answer is that identities need to be vouched for by a trusted third party called a *Certification Authority*. For example, if you want to operate a Web site that needs to perform secure transactions and must be trusted by its users, you need to register with a Certification Authority. The process of registering a Web site is described in Chapter 13.

The Certification Authority will check that you run a reputable organization, and after paying a fee, you will be issued a message called—not surprisingly—a certificate. The certificate identifies you, includes the public key being certified, and adds some extra information that reliably identifies the Certification Authority itself. The certificate testifies that you have proven your identity, and it vouches for the fact that this is your public key. Finally, the certificate is digitally signed by the Certification Authority (using its own private key to encrypt a message digest on the rest of the information). Certificate information is displayed in Chapter 13.

There already are several Certification Authorities in operation.[3] They have well-known public keys that are shipped with browser products such as Netscape and Microsoft Internet Explorer. Banks and credit card issuers are being added to the ranks of Certification Authorities. It is expected to become routine to have certified public keys integrated into your credit cards. Smart cards that do this already exist. As mentioned earlier, there are card devices that you can plug into your personal computer. Software is being integrated into browsers such as Netscape and Microsoft Internet Explorer that steps you through the process of obtaining certification and loading your signature and certificate into your card.

## Secure Applications

Public key encryption and electronic signatures are being used today for secure Web sites and secure electronic mail. These technologies also form the basis for the *Secure Electronic Transactions* (SET) standards, which support credit card purchases and safe business-to-business transactions on the Internet.

## Keeping Up-to-Date

Security is a rapidly evolving area, and new or improved products are appearing at a dizzying rate. In the sections that follow, we will describe some Internet resources that can help you to keep up with the information that you need in order to preserve your site's security.

---

[3]For example, Verisign and GTE.

## CERT Coordination Center

In 1988, a Cornell graduate student named Robert Morris accidentally unleashed a program (dubbed "the Internet worm") that replicated itself across the Internet, generating huge amounts of traffic and blocking access to many parts of the Internet. Various network operations centers struggled to coordinate their information about what was happening, track the source of the problem, and restore normalcy.

The Computer Emergency Response Team (CERT) was formed at Carnegie Mellon University as a result of this event. The CERT Coordination Center continues to be an important central resource for vital security information. For example, at this site you can find out which versions of server application software are most secure (and which have serious flaws).

New security attacks, weaknesses in computer operating systems, and defects in application software are reported to CERT, which posts the information at their Web site and file transfer archive and issues electronic mail bulletins to all interested parties. CERT provides solutions and software patches as soon as they are available. The CERT center also provides security tips and free security tools, which can be accessed via their Web site, or by direct connection to *ftp://info.cert.org/pub/*.

If you wish to join their mailing list, send a message to *cert-advisory-request@cert.org*. More information is available at the CERT Web site, *http://www.cert.org/*.

## The National Computer Security Association (NCSA)

The National Computer Security Association is an independent organization that distributes security information and certifies security products. Its Web site can be accessed at *http://www.ncsa.com/* and contains lists of tested and certified virus checking programs and firewall products. The site also features security news from around the world, such as:

> The June 3 issue of the "London Times" reported that hackers had been paid 400 million pounds sterling in extortion money to keep quiet about having electronically invaded banks, brokerage firms and investment houses in London and New York with "logic bombs."
>
> According to the article, banks chose to give in to the blackmail over concerns that publicity about such attacks could damage consumer confidence in the security of their systems.

NCSA offers a service that performs a test and audit of a Web site. If the site passes the tests, it can display the NCSA certification logo, which builds user confidence in the safety of the site. Certified sites are listed at NCSA's Web server.

# Security Tools

For many years, Unix was the dominant operating system for Internet servers. Unix systems have been pounded in every possible way, and a large number of security tools have been developed that test and monitor Unix security. A good list of these is available at CERT.[4] The Purdue University Computer Operations, Audit, and Security Technology (COAST) site also is a good source for tools. (COAST is discussed in the Reference section at the end of this chapter.) We'll describe a few sample tools below. Computer vendors are starting to incorporate similar tools into new releases of their operating systems.

## CRACK Unix Password Tool

The free CRACK tool tests the passwords at a Unix system. Unix passwords are stored in a file that contains usernames in the clear and passwords in encrypted form. CRACK does not decrypt these passwords. It runs through words in a dictionary, encrypts them using the Unix algorithm, and compares the results to passwords in the password file. This is a useful tool for a Unix administrator who wants to perform an offline check that users have not selected passwords that are easily guessed.

On some Unix systems, it is possible for any logged-in user to copy and view this file. Since CRACK also is freely available to hackers, its existence is a good argument in favor of storing the encrypted passwords in a location that cannot easily be accessed by end-users. The usual way to do this is install *shadow password* software at the Unix computer.

## ScanNT

The ScanNT product performs a password testing scan for NT systems. A demonstration version currently is available. See:

*http://www.omna.com/Yes/AndyBaron/pk.htm*

---

[4]See *ftp://info.cert.org/pub/tech_tips/security_tools.*

## COPS

The Computer Oracle and Password Protection Program, COPS, is a package of auditing routines that check the security of a Unix host. COPS checks whether file access permissions are set appropriately, and watches for suspicious changes in programs that can run with a privileged status. COPS only produces a report. It does not repair problems.

## SATAN

SATAN scans systems from the outside, probing for known security problems. At first, SATAN was viewed with alarm because its point and click user interface and extensive help files made it easy to pinpoint hacker opportunities. Now it is recognized as a very useful tool.

## Internet Security Scanner

The Internet Security Scanner tools are products, not free packages. However you can download test versions of these NT and Unix-based scanners from the ISS Web site, *http://iss.net/*.

# Getting to Know the Enemy

Breaking into computers is considered a form of sport by some and a profession by others. If you want to find out what hackers are up to, you can consult online magazines like *Phrack* and *2600: The Hacker Quarterly*:

*http://freeside.com/phrack.html*

*http://www.2600.com/*

Or, if you prefer to get information in book form, you can buy a paperback such as *Secrets of a Super Hacker* by the Knightmare (Loompanics Unlimited, March 1994).

## REFERENCES

The Computer Operations, Audit, and Security Technology (COAST) project at Purdue University represents the largest dedicated, academic computer security research group in the world. Their Web site provides a treasure trove of information and free security tools. It is located at:

*http://www.cs.purdue.edu/coast/*

If you are concerned about NT security, check the online article at:

*http://www.somarsoft.com/security.htm*

The following books have become a standard part of a security administrator's library:

W. Cheswick and S. Bellovin, *Firewalls and Internet Security* (Addison-Wesley, 1994).

B. Schneier, *Applied Cryptography,* 2nd edition (John Wiley & Sons, 1996). The current versions of the IP security proposed standards are:

RFC 1825 *Security Architecture for the Internet Protocol,* R. Atkinson, August 1995.

RFC 1826 *IP Authentication Header,* R. Atkinson, August 1995.

RFC 1827 *IP Encapsulating Security Payload (ESP),* R. Atkinson, August 1995.

# 11

# Electronic Mail

Electronic mail is on its way to rivaling the telephone as a business tool. With email, you can communicate with any other email user in the world at any time of the day or night. And best of all, once you have paid for your basic Internet connection fee, there is no additional charge for electronic mail, wherever it may be going.

Internet mail would be useful if all that you could do was to exchange text messages with other mail users. But up-to-date email packages also let you tack attachments onto your mail messages. An attachment can be a word processing document, a spreadsheet, a graphic image, a CAD-CAM drawing, a sound file, an executable code—or any other type of data that you wish to send. Mail items often are delivered within minutes, or even seconds, of when they are sent.

If your business is spread across several locations, or if you often need to exchange documents with clients or business associates, then you might be able to save a substantial amount of money by using electronic mail instead of an express package service.

You also can perform business transactions via email. For example, you can set up an order entry form at a World Wide Web server site. After a user fills in the form, an order automatically is sent to you via email. You can also perform business-to-business transactions by sending and receiving secure electronic mail messages, or by sending and receiving formatted Electronic Data Interchange (EDI) messages. Low-cost software and services are available that enable any business to exchange business forms (such as purchase orders) with customers and suppliers across the Internet, and secure electronic mail makes it possible to do this safely.

# Importance of Internet Mail Standards

Internet mail is based on standards. If you have an email product that conforms to these standards, then you can exchange mail with any other conforming Internet mail user. All of the popular TCP/IP application packages conform to these standards.

There are Internet mail standards that describe how mail should be transmitted, such as:

- Simple Mail Transfer Protocol (SMTP): the classic Internet protocol for transmitting mail.
- Extensions to SMTP (ESMTP): an updated version of SMTP that can carry binary information efficiently.

There are Internet mail standards that describe how mail should be formatted, such as:

- Standard for Internet text messages: a simple, classic format for text-based Internet mail.
- Multipurpose Internet Mail Extensions (MIME): an updated format that was designed so that users could add attachments to mail.
- S/MIME: MIME messaging enhanced with electronic signatures and encryption.

# How Internet Email Works

When mail arrives at your business, the postman does not walk around your building, dropping mail onto desktops. Mail is dropped off at a receiving facility, and then is distributed using an internal delivery system.

The same scheme is used for Internet mail delivery. An intermediate system called a *Mail Exchanger* relays mail between the Internet and your

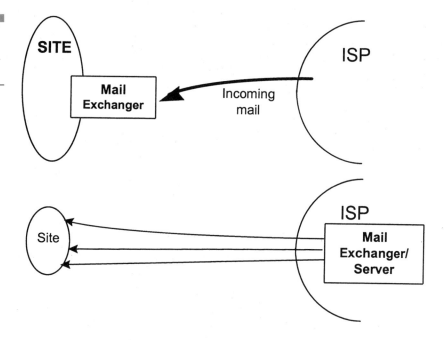

**Figure 11.1**
Relaying electronic
mail via a Mail
Exchanger.

computers. Incoming mail will be delivered to your Mail Exchanger. Similarly, outgoing mail will be forwarded to your Mail Exchanger, and then will be sent on towards its destination.

If you have a large site, you will need to set up a system on your network that will act as your Mail Exchanger. If you only have a few email users, then you may prefer to send and receive mail via a Mail Exchanger/Server that is operated by your ISP. Figure 11.1 illustrates how mail can be relayed to your site via a Mail Exchanger.

## Transmitting Mail to the Internet

Mail products that conform to Internet standards transmit mail via the Simple Mail Transfer Protocol (SMTP) or by its updated version, ESMTP. ESMTP has a number of benefits. The greatest is that it can transmit binaries efficiently.

Figure 11.2 shows two ways that mail might be routed to the Internet from your desktop. An item might be sent to a LAN email server and then forwarded to a Mail Exchanger, or it could be sent directly to the Mail Exchanger.

The outgoing path is chosen when a user configures a desktop mail product. The top part of Figure 11.3 shows how the outgoing path is

**Figure 11.2**
Routing mail to the
Internet.

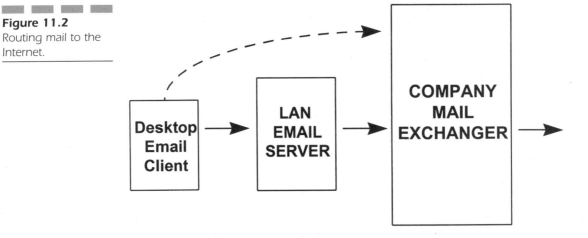

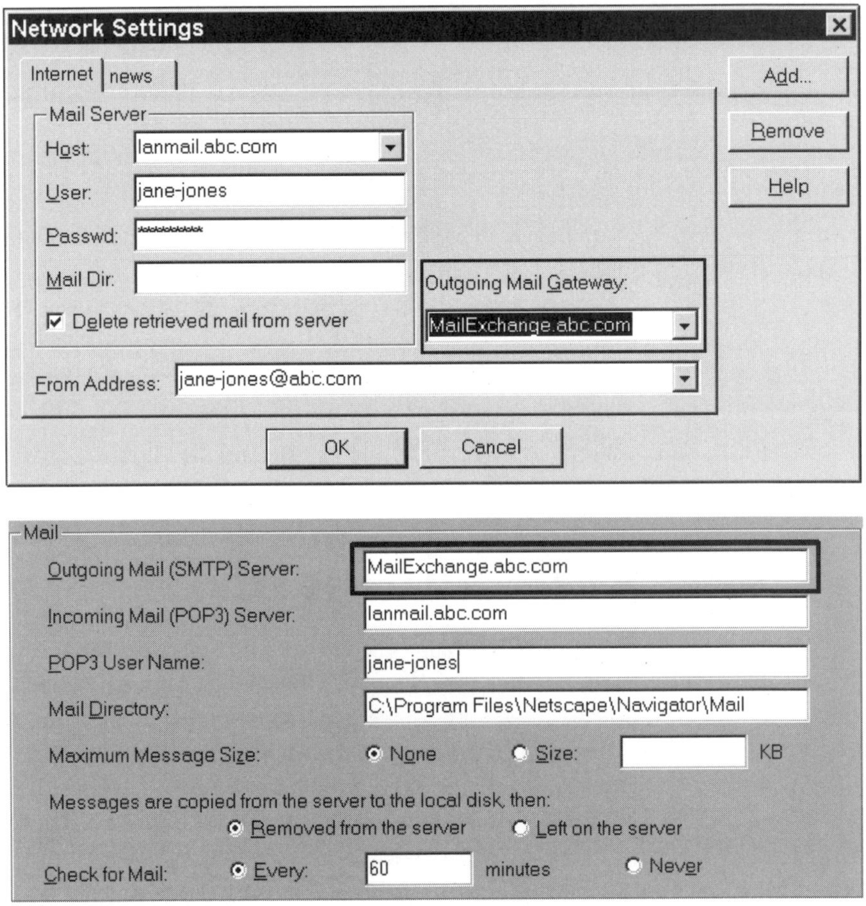

**Figure 11.3**
Configuring outgo-
ing mail for ZMail
PRO or Netscape
Navigator.

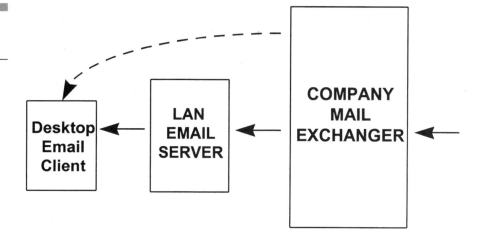

**Figure 11.4**
Routing mail from
the Internet.

configured on mail configuration screens for NetManage ZMail PRO. The lower part of Figure 11.3 shows how the outgoing path is configured for Netscape Navigator mail. The information identifying the system to which outgoing mail should be forwarded has been outlined by black boxes.

## Routing Mail to Your Site

Figure 11.4 shows the path for incoming mail. Mail is transmitted to an organization's Mail Exchanger. From there, it either is delivered directly to desktops, or else is forwarded to a separate LAN email server.

## Role of the Domain Name System

How does a source Mail Exchanger find a destination Mail Exchanger? The Domain Name System plays a crucial role in routing mail to its destination. The names and addresses of Mail Exchangers are listed in the DNS database. To send mail to the ABC company, the source Mail Exchanger looks up the name and address of the Mail Exchanger for *abc.com* in the Domain Name System, and then the mail is transferred. Let's take a look at some sample queries. Before we started the dialogs that follow, we set our query type to MX. The Ashmount *nslookup* program was used to retrieve the following information:

```
Query:Mail Exchange (MX):aetna.com
aetna.com       MX    10       alcfire.AETNA.COM
AETNA.COM       NS    NS.AETNA.COM
AETNA.COM       NS    MIDFIRE1.AETNA.COM
alcfire.AETNA.COM    A    206.25.230.6
NS.AETNA.COM         A    206.25.230.2
MIDFIRE1.AETNA.COM   A    206.156.35.1
Complete: aetna.com
```

The response above shows that Aetna Insurance company receives mail via a Mail Exchanger computer called *alcfire.aetna.com,* whose IP address is 206.25.230.6. Note that the names and address of Aetna's name servers also have been returned. Since we are primarily interested in Mail Exchangers here, we have deleted the names and addresses of the name servers from the next set of responses. Here is information about Cisco Systems Mail Exchangers:

```
Query:Mail Exchange (MX):cisco.com
cisco.com       MX    15       bubbuh.Cisco.COM
cisco.com       MX    10       hubbub.Cisco.COM
cisco.com       MX    10       beasley.Cisco.COM
bubbuh.Cisco.COM     A    198.92.30.35
hubbub.Cisco.COM     A    198.92.30.31
beasley.Cisco.COM    A    171.69.2.135
```

We see that Cisco can receive mail via three different computers. It is not a bad idea to run one or two backup systems in case your main Mail Exchanger is down for maintenance. Which of these systems is Cisco's main exchanger? Actually, two of them are tied for this honor. The numbers in the MX responses (15, 10, and 10 in this case) are called preference numbers. A system with a lower number is preferred. The actual numbers used do not matter at all, just their relative sizes. Hubbub and beasley both have preference numbers equal to 10, and so they have equal status. Next we ask about Verisoft, a security technology company:

```
Query:Mail Exchange (MX):verisoft.com
Authoritative Answer
verisoft.com    MX    50       alterdial.uu.net
verisoft.com    MX    100      mail.uu.net
alterdial.uu.net     A    192.48.96.22
mail.uu.net          A    192.48.96.17
```

Their incoming mail is routed to Mail Exchangers that are operated by their service provider.

# Internet Electronic Mail Identifiers

Anyone who wants to send mail to you needs to know your personal mail identifier. In the old days, Internet mail identifiers had the form:

*userid@hostname*

For example, *brownwm23@hostz23.dept5.abc.com.*

Names like this were hard to remember. They also announced some extraneous information, your username and computer name. This was bad security policy. It also meant that mail IDs might need to be changed fairly often for people in large organizations because they get moved around quite a lot, and their computer names and usernames often are altered in the process.

Today, organizations design and use logical email names that are easy to remember and easy to guess. Forms like the following are popular:

*william-brown@abc.com*

*william_brown@abc.com*

*william.h.brown@abc.com*

There is no hard and fast rule on how to format the first part of the name so although the forms are similar, they are not identical.

# Translating Logical Names

Now, assuming that the ABC company uses the first form above, if I send mail to *william-brown@abc.com,* how is it routed to the recipient?

First of all, the mail is forwarded to an ABC Mail Exchanger. A list of logical names and their actual internal email identifiers is maintained at each Mail Exchanger. When mail addressed to *william-brown@abc.com* arrives at ABC's Mail Exchanger, the name *william-brown* is looked up, the mail is readdressed, header fields containing the email identifier are refor-matted, and then it is sent on. By the way, if ABC uses a proprietary mail product internally, the Mail Exchanger also might have the job of trans-lating the entire message to the internal format. Similarly, a translation from internal format to Internet format would be performed for outgoing mail.

# A Note on Mail Exchanger Lookups

When a user sends mail, the part of the name after the @ symbol is looked up in the Domain Name System using a query of type Mail Exchanger, or MX. If an organization still uses some names like *brownwm23@hostz23. dept5.abc.com*, then extra information in the DNS database is needed in order to route mail to these recipients via the Mail Exchanger. The simplest solution is to include a wild card entry such as:

```
*    MX 10  mail-x.abc.com
```

See Chapter 12 for more details.

# Picking Up Your Mail

Most users today work at desktop systems. Delivery of mail directly to a desktop system usually is not convenient. The system might be turned off or busy with other activities. It is more common to deliver mail to a server. The user connects to the server and picks up incoming mail at a convenient time.

   The Simple Mail Transfer Protocol is not suitable for picking up mail. This task can be carried by either the *Post Office Protocol* (POP) or the *Internet Mail Access Protocol* (IMAP). Figure 11.5 shows the protocols for incoming mail.

**Figure 11.5**
Protocols used to deliver and pick up mail.

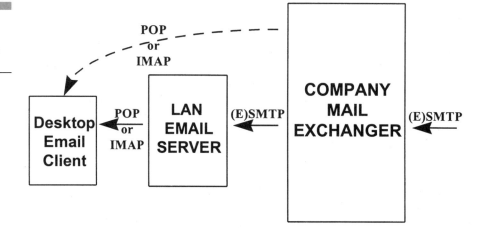

# POP

Today, POP is the most commonly used standards-based method of getting mail from a server.[1] As we noted in Chapter 3, if you have a dial-up Internet account, you pick up your mail from a POP server that is operated by your Service Provider. The POP server often is located at a system that your ISP also uses as a Mail Exchanger. After you have connected your LAN to the Internet, you could install POP server software at your own Mail Exchanger host, or else could set up one or more separate POP LAN servers.

# Configuring a POP Client

Configuring your desktop POP email client is straightforward. The same screens that we showed in Figure 11.3 are used to enter information needed to pick up mail from a POP server. Essential fields are:

■ The name or IP address of the POP server (outlined by black boxes in Figure 11.6).

■ A username and password that will enable you to access the POP server.

You also can identify a desktop directory to be used to store incoming messages and other details, such as how often mail should be retrieved. Configuration for IMAP is similar.

# Setting Up a POP Server

Setting up a POP server also is very straightforward. Free POP server software is available for Unix computers and POP server products are available for other platforms. Be sure to run the current version (POP-3).

Typically, to configure the free POP software at a Unix server, you:

■ Register each POP user with a system username and password. Although they will be listed in the password file, they need not be given interactive access to the computer.

■ Create a mail "spool" directory. Mail will be stored there on a temporary basis.

---

[1]The current version is POP-3.

**Figure 11.6**
Configuring a POP
client.

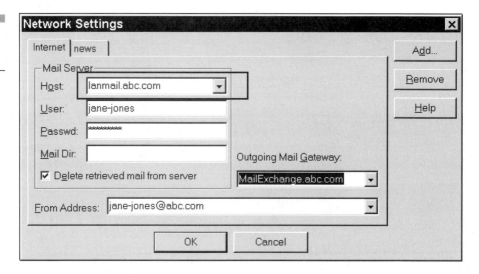

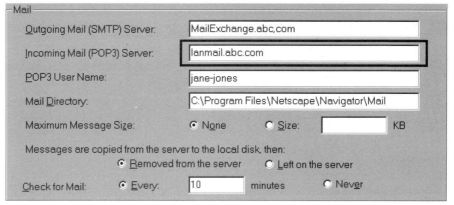

■ Create a mail file for each user. This file must be owned by that user.

■ Schedule the POP server to be started when the computer boots.

Unix products that do not require the POP clients to have system usernames and passwords have appeared on the market. In this case, authorization is handled by the POP server software.

## IMAP

An Internet Mail Access Protocol (IMAP) server is like a POP server with a Ph.D. An IMAP server can act as a long-term repository for some or all of your mail. There is a seamless integration between your desktop and an IMAP server. You can view your mail and store it in a folder at the server. You can move or copy items in either direction between folders at your desktop and folders at the server.

IMAP is just what the doctor ordered for someone who travels a lot and uses a laptop while on the road but uses a desktop in the office. A traveling user often ends up with some mail on the office computer and some on the laptop. If you keep a complete copy of your mail at an IMAP server, you can access all of your mail wherever you may be.

IMAP provides another benefit. An IMAP server that is backed up on a regular basis by a LAN administrator is a safe place to store your mail. IMAP software has been available on the Internet for quite a while, but IMAP products are still fairly new.

# Communicating with Proprietary Mail Systems

Like the telephone system, mail is useful because you can reach a large number of people. It is easy to exchange mail with anyone who has an email package that conforms to TCP/IP standards. However, over the years, many organizations have installed proprietary electronic mail products, or use a proprietary mail online service.

Commercial mail products were not built to conform to Internet standards. Each has its own proprietary format and transmission rules. The vendors of these packages have created "gateway" components that you can install in a mail server or a Mail Exchanger. The gateway software translates between the proprietary product and Internet mail. This enables users of the local, proprietary product to exchange mail with Internet users. Most of the leading products support MIME attachments. Be sure to check that your email system supports MIME.

Integration of Internet mail with the mail product that you like to use every day would be an ideal situation. However, if you have a local mail package that falls short in Internet compatibility, it is not really very difficult to temporarily add a compliant package to your desktop. Both Netscape Navigator and Microsoft Internet Explorer include MIME-compliant mail components. Using an Internet-style package will become more urgent as the use of interworking secure Internet mail products becomes the prevalent mode of electronic mail communication.

Online services such as America Online, CompuServe, and Prodigy each also invented and used their own proprietary electronic mail protocols. The proprietary online services are engaged in a slow process of plugging in conversion capabilities that enable them to support Internet mail functions. All of them can exchange text mail with the Internet. At the time of writing, none of them could handle MIME in a satisfactory manner, although all had promised full support soon. If you are communicating with someone who subscribes to a proprietary service, you will have to check up on its special restrictions.

For some of these services, the only way to send anything other than a text message is to use a conversion tool such as *uuencode,* which can convert a binary file into something that looks like text and can be embedded in a text message. *Uuencode* is a tool that originated in the Unix world and has been ported to other platforms.

# Internet Mail and Security

Mail has had a lot of security problems. The good news is that solutions are making their way into the marketplace. The bad news is that it will take a long time for millions of computers to cut over to more secure software. In the meantime, it is important to be aware of the sources of trouble and take sensible precautions.

# Secure Electronic Mail Client Products

In Chapter 10 we described how Message Digests, public/private keys, symmetric encryption, and certified electronic signatures are being incorporated into Internet communications.

The first widespread use of secure mail was based on a free implementation called Pretty Good Privacy (PGP). PGP, created by Phil Zimmermann, includes all of the security mechanisms discussed in Chapter 10. Recall that PGP was adopted by the InterNIC registration service to improve the authentication of registration updates. PGP, Inc. now markets a secure electronic mail product.

Several major vendors have created electronic mail packages that conform to an interoperable format called S/MIME, which includes certification. S/MIME is expected to become the prevalent form of secure electronic mail, and may become an official standard for electronic mail in the future.

## Screening Electronic Mail

MIME makes electronic mail a lot more useful, but it also can make electronic mail more dangerous. A user may receive an attachment that contains a program that contains viruses or acts as a Trojan Horse. Some sites have installed software that screens incoming electronic mail. This can be helpful, but users still need to be trained to be wary.

## *Sendmail's* Open Door

Mail is delivered to a computer using the venerable Simple Mail Transfer Protocol, SMTP, or its updated version, ESMTP. In either case, the sending computer opens a TCP session to port 25 at the receiving computer. Mail transfer was designed in a very simple-minded way. The sender identifies the mail originator and recipients, and then transmits the contents. All of this is done in a chatty, conversational format. Unfortunately, it is easy for an end user to *telnet* over to port 25 and type in a message.

A mail spoofer can stick any identifier at all in the From: field. The identity of the source computer can be hidden by reporting a fake delivery path. Classic Internet mail offers no assurance that a sender's identity is authentic. This makes the quick adoption of secure electronic mail all the more urgent.

Unfortunately, forged email might be just the beginning of your worries. For years, Unix systems have transmitted and received mail using a program called *Sendmail*. *Sendmail* is big and complicated. Big, complicated programs tend to have security bugs. *Sendmail's* bugs can be dangerous because *Sendmail* runs as a program that has a lot of privileges. It can read and write files anywhere on a computer's disk and it can execute programs. *Sendmail* is vulnerable because an end user can simply *telnet* to the mail port (25) at a system running *Sendmail*, and start talking to *Sendmail*. The user can try various strategies to get control of the computer.

A lot of work has been done on *Sendmail* and many bugs have been removed. However, new ones still pop up occasionally. Why do people continue to use it? *Sendmail* has a lot of very powerful and useful capabilities. It can translate between friendly logical names like *john_jones@abc-corp.com* and actual sender or recipient names. It can translate between just about any types of mail header formats.

# Using SMAP and SMAPD

Fortunately, there is a way that you can enjoy the functionality of *Sendmail* without putting your Mail Exchanger in jeopardy. *Sendmail* is at risk when it receives mail. Fortunately, the job of accepting mail and writing it into a temporary holding file in a spool directory[2] is totally straightforward and several programs have been written to take over that part of *Sendmail's* responsibilities.

For example, the Simple Mail Access Program (SMAP) is part of the free firewall toolkit from Trusted Information Systems. SMAP is a small, safe program that listens at port 25, accepts mail, and stores it in a temporary file. It has a partner program, SMAPD, which periodically checks to see if new mail files have arrived in the spool directory. If desired, SMAPD can be configured to call one or more security screening programs that examine the mail item. For example, it could start a program that checks the mail's attachments for viruses. If the item is clean, then SMAPD passes the mail to *Sendmail,* which can then safely do its job.

# POP and IMAP Weakness

A security flaw in current implementations is that usernames and passwords are passed from a POP client to a POP server in the clear. Furthermore, the current version does not cut off POP login attempts after three tries, so that POP servers are popular targets for hackers who are trying to guess passwords.

If you access an electronic mail server across the Internet, your username and password information might be picked up by an eavesdropper. This is less likely to happen if the mail server is operated by your own ISP and you have dialed directly into your own ISP. It is a danger if you cross intervening networks to reach your mail server. Some ISPs will not accept connection attempts that originate in external networks. This is a good idea. However, keep

---

[2]A spool directory is a computer term meaning a directory used to hold temporary files.

in mind that if you travel, you might have to make a long distance call to pick up your electronic mail.

A hacker with an account on your ISP's network may still try to access mail by guessing user passwords, so it is very important to use complex and hard-to-guess passwords to protect your mail.

There are proposed updates to POP and IMAP that support challenge handshakes or even stronger authentication. This will thwart eavesdroppers. In the meantime, some vendors are offering more secure mail clients and servers.

# Implications of Secure Mail

Once the use of secure, certified electronic mail clients becomes widespread, electronic mail will become an important business tool. It will be used for everything from travel reservations to ordering goods or scheduling services. Organizations will be able to exchange documents, graphic images, and software between sites at negligible cost.

## REFERENCES

Many documents and books have been written relating to electronic mail. A sampling is shown below:

RFC 821 *Simple Mail Transfer Protocol,* J. Postel, September 1, 1982.

RFC 822 *Standard for the Format of ARPA Internet Text Messages,* D. Crocker, September 13, 1982.

RFC 1869 *SMTP Service Extensions,* J. Klensin, N. Freed, M. Rose, E. Stefferud, and D. Crocker, November 6, 1995.

RFC 2045 *Multipurpose Internet Mail Extensions (MIME) Part One: Format of Internet Message Bodies,* N. Freed and N. Borenstein, December 2, 1996.

RFC 2060 *Internet Message Access Protocol—VERSION 4rev1,* M. Crispin, December 4, 1996.

RFC 2095 *IMAP/POP AUTHorize Extension for Simple Challenge/Response,* J. Klensin, R. Catoe, and P. Krumviede, January 30, 1997.

Information about S/MIME can be found at:

*http://www.rsa.com/rsa/S-MIME/home.html*

# 12

# Setting Up Domain Name Servers

We already have seen that name servers play a crucial role in enabling users to connect to Internet servers. In this chapter, we will focus more closely on what lies inside Domain Name System (DNS) databases, and we'll show you how to set up Domain Name Servers on NT and Unix systems.

First we'll get acquainted with the types of information that can be retrieved from a name server. Then we'll look at two implementations—Microsoft's NT DNS server and the popular Unix implementation of the BIND DNS server. Finally, we'll look at how name servers can be set up to function most securely when you connect your site to the Internet.

A lot more can be said about DNS; whole books have been written about it. If your site is very large and will be completely open to the Internet (as is the case with colleges and universities), it is a good idea to look at the material that is referenced at the end of this chapter. In this chapter, we'll focus primarily on two special cases and assume that either you are setting up direct access for users on a single LAN, or else you are going to hide most of your site's computers from the Internet by using proxy firewalls.

# Primary and Secondary Servers

In Chapter 4, we mentioned that you will be required to operate two or more public name servers. One of your servers is special. It is the place that your database information is entered and updated and is called your *primary Domain Name Server*. All of your other DNS servers are called *secondary Domain Name Servers*. A secondary server obtains a copy of the database from the primary.[1]

When a user sends a query to one of your servers, the user neither knows nor cares whether it is a primary or a secondary. You are required to enter database records at one specific server in order to simplify the administration of the Domain Name System and prevent inconsistencies. All of your servers contain authoritative information about your domain.

# Types of Server Information

Before we take a look at what it takes to set up name servers, let's perform a short demonstration that illustrates the types of information stored at DNS servers. We'll login to a Sun Microsystems Unix computer at Yale and talk to a local name server using the Unix *nslookup* tool.[2] *Nslookup* sends queries to a name server—but how does *nslookup* find a local server? On Unix systems, this information is listed in a file called */etc/resolv.conf*.[3] Here is the */etc/resolv.conf* file at the local host:

```
domain CS.YALE.EDU⁴
nameserver 128.36.0.36
nameserver 128.36.0.1
nameserver 128.36.0.3
```

Now we are ready to start *nslookup*:

```
> nslookup
Default Server: DEPT-GW.CS.YALE.EDU
Address: 128.36.0.36
```

---

[1]Later on, we will be more precise in defining primary and secondary functionality.

[2]Some people prefer to use a function called *dig* instead.

[3]On Windows and Macintosh systems, the addresses of DNS servers are entered into a TCP/IP configuration screen.

[4]If a user enters an incomplete name, such as "sunshine," then the domain line indicates that this partial name should be extended to *sunshine.cs.yale.edu*.

*Nslookup* immediately announces the server that it is using, which is the first one on the list in */etc/resolv.conf.* (Another server on the list would be tried if that one was not available.) The same server announcement is printed every time we enter a query. Since repeating this announcement does not add any useful information, we have deleted those lines from the dialogue that follows.

# Internal Address Information

The DNS database enables us to translate names to addresses. We'll illustrate this by asking for the address of a computer named *applejack.cs. yale.edu.*

```
> applejack.cs.yale.edu.
Name:   applejack.cs.yale.edu
Address: 128.36.0.131
```

Note the period at the end of the name. Complete names end with a period that stands for the "root" of the name tree. *Nslookup* would work even if we did not type the period, but it would work inefficiently. See the section entitled "A Tip for Administrators" near the end of this chapter for an explanation.

# Alias (CNAME) Records

You can put nicknames (alias names) into your database. Nickname entries are formally called CNAME (common name) entries. In the example that follows, when we ask for the address of *www.yale.edu,* we discover that this name and *www.cis.yale.edu* both are nicknames for a computer named *elsinore.cis.yale.edu.*

```
> www.yale.edu.
Name:   elsinore.cis.yale.edu
Address: 130.132.143.21
Aliases: www.yale.edu, www.cis.yale.edu
```

You might find it convenient to use a nickname for your World Wide Web server. Someday, you might wish to move your WWW service to a larger computer, or a computer that has a faster connection to the Internet, or even into your Service Provider's network. After you have copied your web pages to the new computer, you can switch users over to the larger computer by making a simple change to a nickname record in your DNS database.

# External Address Information

Next we look up the address of a site external to Yale—the McGraw-Hill World Wide Web site.

```
> www.mcgraw-hill.com.
Name:  www.mcgraw-hill.com
Address: 204.151.55.44
```

Where did this answer come from? Our name server retrieved it from a remote name server that was set up to respond to queries about McGraw-Hill computers.

# Stored Non-authoritative Answers

Below we will enter the same query again. This time the answer does not come from the remote McGraw-Hill server. The first time that our local server (*dept-gw.cs.yale.edu*) obtained the answer, it saved the information so that it would not have to look it up again. Our local server will signal the fact that the answer was stored locally by stating that it has returned a non-authoritative answer.

```
> www.mcgraw-hill.com.
Non-authoritative answer:
Name:  www.mcgraw-hill.com
Address: 204.151.55.44
```

Saving the answers from previous queries enables the local server to answer queries very quickly and also cuts down on Internet traffic. The answers will not be saved forever because DNS entries at the remote site might be changed. The stored answers will be discarded after a time-out period.

Before our server could perform the initial lookup of *www.mcgraw-hill.com*, it had to find out where the McGraw-Hill name servers are located. We'll enter a query that asks specifically where the McGraw-Hill name servers are located by switching to a query of type *name-server* (*ns*).

This answer also is taken from non-authoritative information stored at our local server. Information identifying the McGraw-Hill name servers was retrieved and saved by our own server when it handled the original *www.mc-graw-hill.com* query.

```
> set type=ns
mcgraw-hill.com.

Non-authoritative answer:
mcgraw-hill.com nameserver = NS.ANS.NET
mcgraw-hill.com nameserver = NIS.ANS.NET

Authoritative answers can be found from:
NS.ANS.NET    internet address = 192.103.63.100
NIS.ANS.NET    internet address = 147.225.1.2
```

By the way, note that McGraw-Hill's public DNS database actually is located at computers run by its Internet Service Provider. This is perfectly acceptable. In fact, doing this helps to make McGraw-Hill's own network more secure.

# Using the Root to Find Remote Domain Name Servers

The next question is, how did our server locate the McGraw-Hill name servers? To do this, it sent a request to an ultimate authority—one of the copies of the root database. The root database contains the locations of second-level domains under COM, ORG, EDU, GOV, NET, etc., and also contains pointers to the sites that act as roots for domains under MIL, US, UK, DE, etc.

But how did our local name server know where the root servers are located? This is the end of the line. Our server knows about these root servers because the InterNIC lists them in a file:

*ftp://ftp.rs.internic.net/domain/named.root*

A copy of the current version of this file is stored in every name server host. This root information provides a starting point for remote queries.

We now have traced the full path to the information in our original query. To review, to look up the address corresponding to a name such as *www.mc-graw-hill.com*, our local name server:

■ Sent a query to a root server requesting the location of the McGraw-Hill name servers.

■ Sent a query to one of the McGraw-Hill name servers that retrieved the address of the host.

## Viewing the Root Servers

In the query below, we ask for the list of root servers by entering *"."*, which stands for the root. The response is non-authoritative because it is taken from information that is stored at our local server, *dept-gw.cs.yale.edu.*

```
>  .

Non-authoritative answer:
(root)  nameserver = D.ROOT-SERVERS.NET
(root)  nameserver = E.ROOT-SERVERS.NET
(root)  nameserver = I.ROOT-SERVERS.NET
(root)  nameserver = F.ROOT-SERVERS.NET
(root)  nameserver = G.ROOT-SERVERS.NET
(root)  nameserver = A.ROOT-SERVERS.NET
(root)  nameserver = H.ROOT-SERVERS.NET
(root)  nameserver = B.ROOT-SERVERS.NET
(root)  nameserver = C.ROOT-SERVERS.NET

Authoritative answers can be found from:
D.ROOT-SERVERS.NET    internet address = 128.8.10.90
E.ROOT-SERVERS.NET    internet address = 192.203.230.10
I.ROOT-SERVERS.NET    internet address = 192.36.148.17
F.ROOT-SERVERS.NET    internet address = 192.5.5.241
G.ROOT-SERVERS.NET    internet address = 192.112.36.4
A.ROOT-SERVERS.NET    internet address = 198.41.0.4
H.ROOT-SERVERS.NET    internet address = 128.63.2.53
B.ROOT-SERVERS.NET    internet address = 128.9.0.107
C.ROOT-SERVERS.NET    internet address = 192.33.4.12
```

## The InterNIC *named.root* File

The InterNIC file *named.root,* stripped of all of its comments, is displayed in Figure 12.1. Recall that a period ( . ) means *root.* We will discuss the details of the record format later. For now, note that there are thirteen records that list the names of root name servers. For each of these, there is a record that identifies the server's IP address. The number 3600000 is a timeout expressed in seconds. It causes the local server to retrieve an updated list of root servers every 1000 hours.

## Mail Exchanger Records

Now let's return to our exploration of Domain Name System databases, and look at some more query examples. Below, we change to a query of

**Figure 12.1**

The InterNIC
*named.root* file.

```
.                       3600000  IN  NS   A.ROOT-SERVERS.NET.
A.ROOT-SERVERS.NET.     3600000      A    198.41.0.4
.                       3600000      NS   B.ROOT-SERVERS.NET.
B.ROOT-SERVERS.NET.     3600000      A    128.9.0.107
.                       3600000      NS   C.ROOT-SERVERS.NET.
C.ROOT-SERVERS.NET.     3600000      A    192.33.4.12
.                       3600000      NS   D.ROOT-SERVERS.NET.
D.ROOT-SERVERS.NET.     3600000      A    128.8.10.90
.                       3600000      NS   E.ROOT-SERVERS.NET.
E.ROOT-SERVERS.NET.     3600000      A    192.203.230.10
.                       3600000      NS   F.ROOT-SERVERS.NET.
F.ROOT-SERVERS.NET.     3600000      A    192.5.5.241
.                       3600000      NS   G.ROOT-SERVERS.NET.
G.ROOT-SERVERS.NET.     3600000      A    192.112.36.4
.                       3600000      NS   H.ROOT-SERVERS.NET.
H.ROOT-SERVERS.NET.     3600000      A    128.63.2.53
.                       3600000      NS   I.ROOT-SERVERS.NET.
I.ROOT-SERVERS.NET.     3600000      A    192.36.148.17
.                       3600000      NS   J.ROOT-SERVERS.NET.
J.ROOT-SERVERS.NET.     3600000      A    198.41.0.10
.                       3600000      NS   K.ROOT-SERVERS.NET.
K.ROOT-SERVERS.NET.     3600000      A    193.0.14.129
.                       3600000      NS   L.ROOT-SERVERS.NET.
L.ROOT-SERVERS.NET.     3600000      A    198.32.64.12
.                       3600000      NS   M.ROOT-SERVERS.NET.
M.ROOT-SERVERS.NET.     3600000      A    198.32.65.12
```

type Mail Exchanger, and ask for information about the McGraw-Hill Mail Exchanger:

```
> set type=mx
> mcgraw-hill.com.
mcgraw-hill.com preference = 20, mail exchanger =
interlock.mgh.com
mcgraw-hill.com nameserver = nis.ans.net
mcgraw-hill.com nameserver = ns.ans.net
interlock.mgh.com    internet address = 152.159.1.2
nis.ans.net    internet address = 147.225.1.2
ns.ans.net    internet address = 192.103.63.100
```

# Text and Other Comment Records

A DNS administrator can embed comments into a database and allow these to be retrieved. For example:

```
> set type=txt
> fhp.jvnc.net.
fhp.jvnc.net    text="First Health Plan, Red Bank, NJ"
```

Actually, three types of comment records are used:

txt        Any text.

hinfo      Host information. A description of the system's hardware and
           operating system.

Wks        Well-known services. A list of services (e.g., WWW or file
           transfer) offered at a system.

These comment records are not used for anything, except perhaps to
remind a local administrator of some information. It may be handy to
plant some reminder messages into a private local server, but announcing the hardware, operating systems, and services available at your systems just makes a hacker's job easier. Most administrators agree that it
is not a good idea to put these records into a public DNS database. An
administrator can easily store comment information into a personal desktop database.

There is one comment record type that currently is experimental and may
turn out to be quite useful. An *rp* (responsible person) record reports the
email address of the person who is responsible for a particular computer. This
would be helpful if users wanted to report some problem associated with that
computer.

# Translating Addresses to Names

Next, we will perform reverse lookups that translate addresses to names. First
we will show you how to do this, and then we will explain it. To perform an
address-to-name query with *nslookup*, set the type to *ptr* (pointer). Below, we
look up the fact that the computer whose address is 204.151.55.44 is *www.mc-
graw-hill.com*.

```
> set type=ptr
> 204.151.55.44
44.55.151.204.in-addr.arpa    name = www.mcgraw-hill.com
55.151.204.in-addr.arpa       nameserver = knock.aa.ans.net
55.151.204.in-addr.arpa       nameserver = dns2-a.ans.net
knock.aa.ans.net              internet address = 198.83.21.10
dns2-a.ans.net                internet address = 198.83.47.29
```

Look carefully at the first line of the output. To get the answer, *nslookup*
had to reverse the order of the numbers in the address and tack on the labels
*in-addr.arpa*. That is, *nslookup* actually sent a query that asked for the name

corresponding to *44.55.151.204.in-addr.arpa.*[5] This odd-looking format actually reflects the under-the-covers structure of the database.

# The Address Subtree

There is a separate Domain Name System subtree structure that is used for addresses. The tree is shown in Figure 12.2. The top name, *in-addr.arpa.*, is a holdover from the ancient history of the Internet and stands for "ARPA Internet addresses." At first glance, the rest of the tree looks upside down. But let's take a closer look. In the name:

*www.mcgraw-hill.com*

the most general label, *com*, is at the top of the naming tree. By convention, names are written with the most general label at the right end.

Now look at the address:

204.151.55.44

the most general label is 204. It must be at the top of the address tree. By convention, the most general label in an address happens to be written at the left end. Note that when you write out address labels from the bottom to top,

---

[5]Earlier versions of *nslookup* forced the end user to enter address queries in this tortured format.

**Figure 12.2**
The Internet address tree.

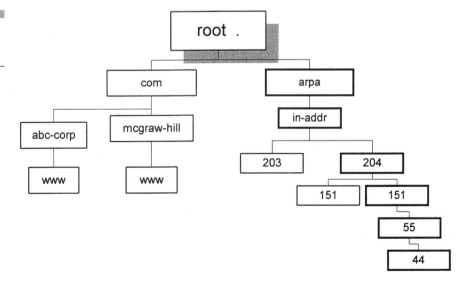

you get the strange looking syntax that appeared in the *nslookup* response: *44.55.151.204.in-addr.arpa.*

# Zones

Imagine what would happen if there were one central agency that was responsible for the entire Internet Domain Name System database. Suppose that every time you added one host to the Internet or moved a system to a different part of your network, you had to send a form requesting that the database be updated.

The reason that the Internet works is that this is not the way that things are done. Instead, the DNS database is broken into small, manageable chunks called *zones*. A zone is a chunk of DNS information about a contiguous part of the name or address tree. A zone is under the control of an administrator. The administrator makes zone information available to the world by entering it into a zone file that is then loaded into a DNS database.

If you are the administrator of a site that will be connected to the Internet, you are responsible for setting up a primary server at which you enter your zone information—the names and addresses of computers in your network that will communicate with other Internet computers. Because this server is so important, you must provide one or more backup servers (called secondary servers) in case the primary crashes or gets too busy to handle the entire load of queries.[6]

Recall that a registration form tells the InterNIC the names and addresses of these servers. The InterNIC will add this information to its *root* list, which contains pointers to DNS servers all over the world.

In the sections that follow, we are going to step through the detailed setup of simple Domain Name Servers for a LAN that will be connected to the Internet. Systems on this LAN have names that end in *mgt.com*. Their addresses belong to the Class C network 198.207.177. The bulk of the work consists of preparing two zone files:

- The *mgt.com* zone file contains address information about the computers whose names end in *mgt.com*, and identifies the Mail Exchanger for *mgt.com*.

- The *177.207.198.in-addr.arpa* zone file maps the addresses on LAN 198.207.177 to names.

---

[6]For a fee, your ISP will set up and run these servers for you, but you are still responsible for making sure that the database information is timely and correct.

# Setting Up Microsoft NT DNS Servers

Microsoft NT Server version 4 includes a Domain Name Server that is very easy to set up. To start out, we will use the Microsoft application to set up primary and secondary name servers at NT systems. Then we will be ready to configure primary and secondary servers at Unix systems.

If you are just learning to set up name servers, it is worth reading the NT example even if you are going to use Unix servers. The NT configuration process is equivalent to the Unix configuration process but automates some of the work so that it is a lot simpler.

# Configuring a Microsoft NT Primary DNS Server

The primary NT server will be set up at a system called *ntserver.mgt.com*. To create the NT server, choose:

Start/Programs/Administrative Tools (Common)/DNS Manager

We will define a new server to operate at the local address, which is 198.207.177.32.

As we see in Figure 12.3, an item called *cache* appears automatically. This contains a list of root servers. Microsoft included records from the Registration InterNIC's current *named.root* file when it shipped the NT operating system software. If you wish, you can download a current *named.root* and, if there are any changes, you can enter them manually. However, this is not really necessary, because when your system boots, it will connect to one of the preconfigured root servers and get an up-to-date list.

In Figure 12.3, we also see that there are three dummy address zones that are automatically created for the NT server: *127.in-addr.arpa*, *0.in-addr.arpa*, and *255.in-addr.arpa*. Systems cannot be given IP addresses that start with 127 or 0, and addresses starting with 255 currently cannot be used. These dummy zones are included so that if the server receives a lookup for an address starting with one of these numbers, the server will not go chasing off to a root server looking for an impossible address.

**Figure 12.3**
Initializing a DNS
server at an NT
system.

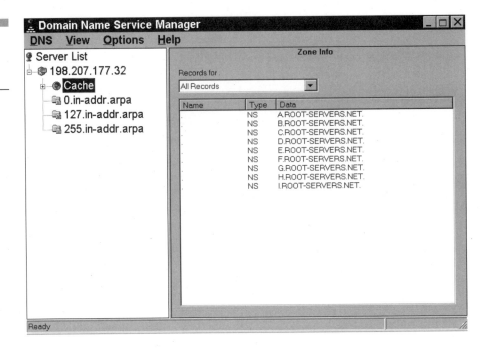

## Adding the *mgt.com* Zone

To add zone data to this server, we position the cursor on our server's address, click the right mouse button, and select **New Zone**. The screen shown in Figure 12.4 appears. Check **Primary** to make this the primary server for that zone.

Next we enter the domain name for this zone, which is *mgt.com*, as shown in Figure 12.5. By default, the records that we enter will be stored in a file called *mgt.com.dns*. This is an ordinary text file. We can change the name in this field to any filename that we like, but we might as well stick with the default.

Three records are created automatically and the most important information from each record is displayed in a window:

```
mgt.com      SOA     ntserver.mgt.com, Admin.mgt.com
mgt.com      NS      ntserver.mgt.com
ntserver     A       198.207.177.32
```

Each record has a type, identified in the second column above.

■ SOA stands for Start of Authority. This record announces that the primary server for *mgt.com* is *ntserver.mgt.com*, and that any problems

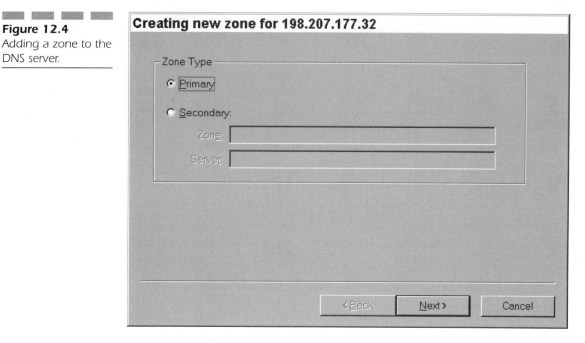

**Figure 12.4**
Adding a zone to the
DNS server.

**Creating new zone for 198.207.177.32**

Zone Type
- ⊙ Primary
- ○ Secondary:
  - Zone:
  - Server:

< Back    Next >    Cancel

**Figure 12.5**
Identifying the name
of the zone file.

**Creating new zone for 198.207.177.32**

Zone Info

Zone Name:    mgt.com

Zone File:    mgt.com.dns

Enter the name of the zone and a name for its database.

< Back    Next >    Cancel

should be reported to email address *Admin@mgt.com*. By convention, the first period in the record has to be changed to an @ to derive the email address.[7]

■ The NS (Name Server) record identifies this computer, *ntserver.mgt.com,* as a name server for the domain *mgt.com*.

■ The record of type A (Address) provides the address of this server.

# Adding the *177.207.198. in-addr.arpa* Zone

We will have to enter the rest of the records manually. But before we do, let's create a new primary zone for our network address, 198.207.177 (see Figure 12.6).

Note that we had to enter our network number in reverse order, followed by *.in-addr.arpa*. Two records are created automatically, and the most important information from each record is displayed in a window:

---

[7]When an @ appears in a DNS file, it has a special meaning and gets translated to something else. Hence the usual @ that separates an ID from the rest of an email address can't appear here.

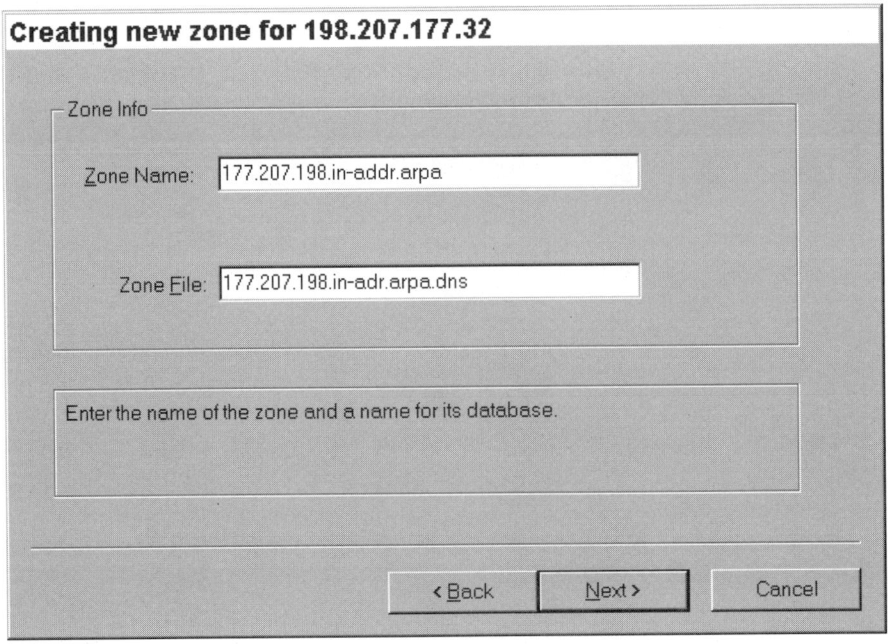

**Figure 12.6**

Creating an address zone.

Creating new zone for 198.207.177.32

Zone Info

Zone Name: 177.207.198.in-addr.arpa

Zone File: 177.207.198.in-adr.arpa.dns

Enter the name of the zone and a name for its database.

< Back    Next >    Cancel

```
177.207.198          SOA      ntserver.mgt.com, Admin.mgt.com
32.177.207.198       NS       ntserver.mgt.com
```

*Ntserver.mgt.com* is the primary name server for the new zone, *177.207.198. in-addr.arpa.*

# Entering Resource Records

The records in a zone file formally are called *Resource Records*. Figure 12.7 shows how we enter a record into zone *mgt.com* that maps the name *compaq* to an address. The reason that we wanted to create the *177.207.198.in-addr.arpa* zone file before entering the rest of the *mgt.com* records is that this NT tool will automatically create corresponding address-to-name records for us in the *177.207.198.in-addr.arpa* zone file. This saves us some work and prevents us from making a typing error or forgetting to include some of the reverse records.

Note the Time To Live (TTL) value in the left lower box. When a remote server asks for this record, the local server will send the TTL back as part of

**Figure 12.7**
Entering Resource Records.

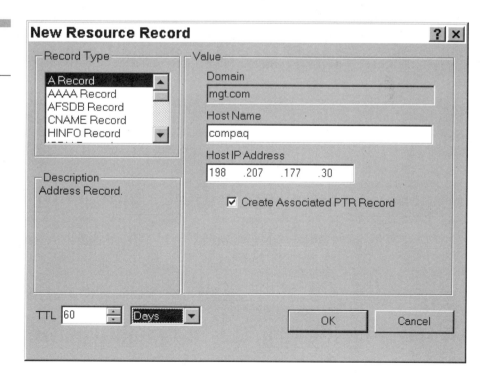

the response. This tells the requester how long the answer may safely be stored at its system. If I were going to change the IP address of this computer next week, I would change the TTL to a smaller value so that the old information would get dumped out of remote servers in a timely manner.

Now we are ready to enter the rest of our records. Clearly, Web, mail, and file transfer servers need to be listed, because users will want to access them by name. It might come as a surprise to learn that client systems on our LAN that have visible IP addresses and wish to communicate with Internet servers also need to be listed in our public database. The reason is that some Internet servers perform a weak form of client validation by:

■ Using the IP address of an incoming client to look up the client's name.

■ Looking up the name and making sure that the resulting address matches the original address.[8]

Many servers actually turn down clients for whom these lookups fail. This means that each visible client needs to have a complete and consistent set of entries. If a proxy is used to access Internet servers, then the only client system that would need to be listed would be the proxy.

## Loading the Data

By default, the NT DNS server starts up at boot time. After we have finished entering all of our information, if we click on

DNS/Update Server Data Files

the new data will be loaded into the active database. The application is now ready to be queried for the information that we just entered.

## Secondary NT Server

Next, we will create a secondary server at another NT system called *toshnt.mgt.com* at 198.207.177.41. The steps are:

■ Choose DNS/New Server, and enter the address of the primary that already is running, 198.207.177.32.

---

[8]This procedure is a bad performance drag on a server.

**Figure 12.8**
Creating a secondary zone at an NT DNS server.

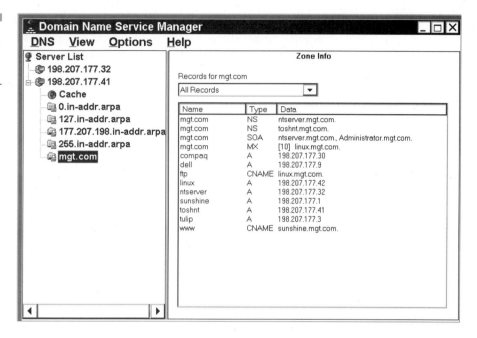

- Choose **DNS/New Server**, and enter the address of this server (*toshnt*), 198.207.177.41.

- Add the zone *mgt.com,* click "secondary," and provide the address of the primary server.

- Add the zone *177.207.198.in-addr.arpa,* click "secondary," and provide the address of the primary server.

The secondary immediately connects to the primary and downloads the records for these two zones. Figure 12.8 shows the result. All of the records now are available at the secondary server, which is ready to respond to queries. From now on, the secondary will connect to the primary at regular intervals and will perform a fresh zone transfer for any zone that has been updated.

# Loading Multiple Zones

In the example above, we set up servers that provided data about one small LAN. However, a name server can be loaded up with data for lots of zones and will happily answer queries about any of this information. The biggest name servers are operated by Internet Service Providers who routinely provide DNS service on behalf of their customers—for a fee. To do this, they load separate zone files for each of their customers into a primary server, and use one or two of their other servers as secondaries.

# Primary and Secondary Servers for a Zone

Strictly speaking, it is not correct to say that a system is a "primary server" or a "secondary server." Any DNS server will do whatever you ask it to do. If you create some zone files at a server and tell the server that it is primary for those zones, the server will load the files, answer queries, and transfer the zone information to its secondaries when it is asked to do so. If you tell the server that it is a secondary for some other zones, then it will dutifully contact the primaries that you have identified, request the zone data, and answer queries.

Thus, the correct terminology is to say that a DNS server is primary (or secondary) for a particular zone. People are rarely this precise in their speech, and usually understand what primary or secondary server really means from context.

# Viewing the *mgt.com.* Zone File

Now we can leave the pretty user interface and take a look at the actual zone files that were created by the NT DNS application. These zone files have exactly the same format as the zone files that are used with the Unix BIND program. Let's examine the *mgt.com.dns* file, which is listed in Figure 12.9. The information in the file is more detailed than the summary data displayed by the NT DNS user interface. Comments have been inserted, preceded by the semicolon ( ; ) character. The "IN" in each record stands for "Internet." The Domain Name System was designed so that it could be used with multiple protocol families.

The SOA, NS, and MX records start with @, meaning "the current origin." This is the special use of the @ symbol that we mentioned earlier. NT knows what the current origin is because there is an entry in the NT system registry that states that this file contains records for names in the *mgt.com.* zone. It

**Figure 12.9**

File for the *mgt.com* zone.

```
;
;   Database file mgt.com.dns for mgt.com zone.
;       Zone version:  41
;

@               IN  SOA  ntserver.mgt.com.   Admin.mgt.com.      (
                        4               ; serial number
                        3600            ; refresh
                        600             ; retry
                        86400           ; expire
                        3600        )   ; minimum TTL

;
;   Zone NS records
;

@                               IN      NS      ntserver
@                               IN      NS      toshnt


;
;   Zone records
;
@                   5184000     IN      MX      10      linux.mgt.com.
compaq              5184000     IN      A       198.207.177.30
dell                5184000     IN      A       198.207.177.9
ftp                             IN      CNAME   linux
linux               5184000     IN      A       198.207.177.42
ntserver            5184000     IN      A       198.207.177.32
sunshine            5184000     IN      A       198.207.177.1
toshnt              5184000     IN      A       198.207.177.41
tulip               5184000     IN      A       198.207.177.3
www                             IN      CNAME   sunshine
```

also is correct to be explicit and write *mgt.com.* instead of @ at the start of the SOA and NS records.

There are a lot of numeric parameters in the SOA record that were not included in the earlier display. All but one of these numbers provide instructions to secondary servers telling them how to behave. When a secondary server downloads a zone, it looks at these numbers to discover information such as how often it should reconnect to the primary. Default numbers were automatically inserted into the SOA record, but there is an NT screen that enables you to edit these numbers.

The first number is a serial number. Every time a database zone file is updated at the primary, the serial number must be increased. Every time that a secondary for a zone connects to the primary, it checks this number. If there have been no new updates since the last visit, then the serial number will not have changed, and the secondary knows that it does not have to refresh its copy of this zone.

The remaining numbers express timeouts in seconds. The timeouts in this example are:

| | |
|---|---|
| Refresh: | Secondaries should connect to the primary at intervals of 3600 seconds (1 hour) and check whether they need to refresh this zone. |
| Retry: | If a secondary is unable to connect at the scheduled time, it should try again after 10 minutes. |
| Expire: | If the secondary has been unable to connect for 24 hours, it should consider its data to have expired and should stop answering queries. |
| Minimum TTL: | This is a misnomer. It actually is a default Time To Live value, used for entries that have not been assigned a specific TTL value. |

For the remaining records, the main difference between the file format and the NT user interface screen is the fact that the Time To Live values are displayed, and the "IN" keyword appears.

All of the names mentioned in this zone file that do not end with a "." will automatically be completed by adding *mgt.com.* at the end. For example, *compaq* translates to *compaq.mgt.com.*

Recall that CNAME records are used to assign nicknames. There is a public file transfer server running at *linux.mgt.com* and a World Wide Web server running at *sunshine.mgt.com.* Users will have an easier time finding these services if we assign the nicknames *ftp.mgt.com* and *www.mgt.com* to these computers.

One final observation: Note that NT has nicely sorted our zone records alphabetically.

# Viewing the *177.207.198. in-addr.arpa* Zone File

The file for the address zone *177.207.198.in-addr.arpa* is similar (see Figure 12.10). After the SOA and NS records, address-to-name mappings are listed in PTR (pointer) records. The complete address of each system is obtained by appending .177.207.198 to each of the numbers in the left column and then reversing the order. For example, 1 corresponds to 198.207.177.1.

**Figure 12.10**
File for the
*177.207.198.in-addr.arpa* zone.

```
;
;   Database file 177.207.198.in-addr.arpa.dns for
177.207.198.in-addr.arpa
;   zone.
;        Zone version:  31
;

@         IN    SOA    ntserver.mgt.com.    Admin.mgt.com.
                       3              ; serial number
                       3600           ; refresh
                       600            ; retry
                       86400          ; expire
                       3600         ) ; minimum TTL


;
;   Zone NS records
;

@                               IN    NS     ntserver.mgt.com.
                                IN    NS     toshnt.mgt.com.


;
;   Zone records
;

1                               IN    PTR    sunshine.mgt.com.
3                               IN    PTR    tulip.mgt.com.
9                               IN    PTR    dell.mgt.com.
30                              IN    PTR    compaq.mgt.com.
32              5184000         IN    PTR    ntserver.mgt.com.
41              5184000         IN    PTR    toshnt.mgt.com.
42                              IN    PTR    linux.mgt.com.
```

## Summary of DNS Files

The files that were needed for our DNS server are summarized in Table 12.1 at the top of the next page.

## Multiple Address and Name Zones

Today it is common for sites to have several network numbers; for example, you might have been given a block consisting of four Class C addresses. In this case, you need to add an *in-addr.arpa* zone file for each of your addresses.

**TABLE 12.1**

Files Needed for
the NT DNS Server

| File | Description |
| --- | --- |
| Cache | A file containing the list of root servers. |
| *mgt.com.* zone | A file that maps names to addresses, identifies mail exchangers, defines nicknames, etc. |
| *177.207.198.in-addr-arpa.* zone | A file that maps addresses to names. |
| *127.in-addr.arpa.* zone<br>*localhost*<br>*0.in-addr.arpa*<br>*255.in-addr.arpa* | Files that handle lookups for reserved addresses and prevent the server from going off to the root to look for these addresses. |

You also might have more than one naming zone. For example, McGraw-Hill owns both *mcgraw-hill.com* and *mgh.com*. A separate zone file is needed for each name.

# Keeping Track of the Files

Finally, there is one last piece of configuration data that is needed to pull all of this information together, namely, a list of the filenames of all of the files whose entries will be part of this server's database. In the Microsoft NT server implementation, this list is generated automatically when you enter the data for your various zones, and the list is stored in the NT registry. In the Unix BIND implementation that we will examine next, this list is kept in a separate boot file.

# The BIND Implementation

For many years, the Berkeley Internet Domain (BIND) implementation has been the favorite choice for Internet Domain Name Servers and Unix has been the favored operating system. BIND was originally written at the University of California at Berkeley under a grant from the U.S. government. As a result, the software has always been free. Versions from 4.9.3 onward have been sponsored by the Internet Software Consortium, a nonprofit organization that maintains free BIND, Dynamic Host Configuration Protocol (DHCP), and Internet News (INN) software. At the time of writing, the most current Unix BIND version was 4.9.5.

BIND has been ported to other operating systems. In fact, there now is a free NT version of BIND that you can use instead of the application that is provided by Microsoft.

The BIND DNS server program is called *named* (pronounced "name-d") and usually is located in the */etc* directory (*/etc/named*). As we shall see later, a configuration file called */etc/named.boot* lists all of the zone files that contain information for the database.[9]

# Setting Up Unix BIND Servers

Suppose that instead of using our NT servers, we wanted to use a Unix computer called *sunshine.mgt.com* as our primary server, and a Linux computer called *linux.mgt.com* as our secondary server. *Sunshine* has address 198.207.177.1 and *linux* has address 198.207.177.42.

By the way, we could have used an NT system as one server and a Unix system as the other. All DNS servers behave in a standard way, so you can use any platforms that you like for your servers.

# Configuring a Primary BIND Server

To set up a server that is primary for these zones, we will need all of the files that were listed in Table 12.1. Fortunately, we can reuse the files that were created for our NT server. We can edit the files that were displayed in Figures 12.9 and 12.10 so that the names of the Unix servers appear in the appropriate records.

Figure 12.11 shows a modified *mgt.com* zone file. We could have just replaced the name server information, but we have made a few other changes to show some alternative formats that are valid.[10] Specifically:

■ The "IN" keyword actually is only required in the first record, and has been omitted from the rest.

---

[9]The command that starts *named* could identify a different boot file.

[10]The formatting rules support quite a few different ways to express the same information. Many people find this confusing.

**Figure 12.11**

The *mgt.com* zone
file at *sunshine*.

```
;
;  Database file mgt.com.dns for mgt.com zone.
;

mgt.com.   IN SOA    sunshine.mgt.com.   Admin.mgt.com. (
                     4            ; serial number
                     3600         ; refresh
                     600          ; retry
                     86400        ; expire
                     3600       ) ; minimum TTL


;
;   Zone NS records
;

mgt.com.                    NS      sunshine.mgt.com.
mgt.com.                    NS      linux.mgt.com.


;
;   Zone records
;
mgt.com.          5184000   MX   10 linux.mgt.com.
compaq            5184000   A      198.207.177.30
dell              5184000   A      198.207.177.9
ftp                         CNAME  linux.mgt.com.
linux             5184000   A      198.207.177.42
ntserver          5184000   A      198.207.177.32
sunshine          5184000   A      198.207.177.1
toshnt            5184000   A      198.207.177.41
tulip             5184000   A      198.207.177.3
www                         CNAME  sunshine.mgt.com.
```

■ We have filled in the name of the zone instead of using @ characters.

■ In the CNAME records, we have written out the full names instead of just the first part.

Figure 12.12 shows the modified zone file for *177.207.198.in-addr.arpa*. The format of the file for the dummy domain *0.0.127.in-addr.arpa* is shown in Figure 12.13. This file includes an SOA record, an NS record, and an entry that maps address 127.0.0.1 to the name *localhost*.

Similar dummy files may be used for *localhost, 0.in-addr.arpa,* and *255.in-addr.arpa*. However, check your version of BIND. It might generate all of the needed dummy zones and entries automatically.

Just as before, we can give our files any names that we like. On Unix systems, the file containing the list of root servers often is called *root.cache*.

**Figure 12.12**
The *177.207.198.in-addr.arpa* zone file at *sunshine.*

```
;
;   Database file 177.207.198.in-addr.arpa.dns for
;   177.207.198.in-addr.arpa
;   zone.
;

mgt.com.    IN  SOA    sunshine.mgt.com.    Admin.mgt.com. (
                    3              ; serial number
                    3600           ; refresh
                    600            ; retry
                    86400          ; expire
                    3600        )  ; minimum TTL


;
;   Zone NS records
;

mgt.com.                      NS    sunshine.mgt.com.
mgt.com.                      NS    linux.mgt.com.


;
;   Zone records
;

1                 5184000     PTR    sunshine.mgt.com.
3                             PTR    tulip.mgt.com.
9                             PTR    dell.mgt.com.
30                            PTR    compaq.mgt.com.
32                            PTR    ntserver.mgt.com.
41                            PTR    toshnt.mgt.com.
42                5184000     PTR    linux.mgt.com.
```

The file that maps address 127.0.0.1 to *localhost* often is called *named.local.* We might as well use these conventional names at our server. We can make up any convenient name pattern that we wish for our zone files. We will use the same names that we used earlier: *mgt.com.dns* and *177.207.198-in-addr.arpa.dns.*

Recall that the BIND Domain Name Server program is called *named.* When *named* starts, it needs to read the list of all of the files that need to be loaded into this server's database. By default, *named* will look for this list in a boot file called */etc/named.boot.* If you wanted to put your list into a file with a different name, then the name of that boot file would have to be provided as a parameter to *named* when it was started.

Whatever you may choose to call it, the boot file contents are straightforward. The boot file identifies the directory in which all of the zone files are

**Figure 12.13**
The *0.0.127.in-addr.arpa* zone file at *sunshine*.

```
;
;   Database file  for 0.0.127.in-addr.arpa zone.
;

0.0.127.in-addr.arpa. SOA  sunshine.mgt.com. Admin.mgt.com.  (
                      10         ; serial number
                      3600       ; refresh
                      600        ; retry
                      86400      ; expire
                      86400     ) ; minimum TTL


;
;   Zone NS record
;
0.0.127.in-addr.arpa.        NS      sunshine.mgt.com.
;
;   Zone record
;
1                            PTR     localhost.
```

located and lists the names of the zone files. You can select your own directory and make up your own naming scheme. Figure 12.14 shows the *named* boot file for our primary server.

We have chosen to put our files into a directory called */usr/local/ adm/named*. Note that the zone names are not terminated with a period in the boot file.

Remember that your server can support as many zones as you like. For example, Internet Service Providers often set up servers that contain information for dozens of their clients. Therefore, these servers hold data for many zones. A separate file has to be created for each zone, and then the name of each file must be added to the boot list at the primary server.

**Figure 12.14**
Boot file for the primary BIND Name Server at *sunshine*.

```
;
;   Boot file for Primary Name server
;
directory       /usr/local/adm/named
;
;               zone                        file
;
primary     mgt.com                     mgt.com.dns
primary     177.207.198.in-addr.arpa    177.207,198.in-
addr.arpa.dns
primary     0.0.127.in-addr.arpa        named.local
cache       .                           root.cache
```

**Figure 12.15**

Boot file for a
secondary BIND
server at *linux*.

```
;
;  Boot file for Secondary Name server
;
directory       /usr/local/adm/named
;
;               zone                        Primary Server
;
secondary       mgt.com                     198.207.177.1
secondary       177.207.198.in-addr.arpa    198.207.177.1
;
;                zone                        file
;
primary         0.0.127.in-addr.arpa        named.local
cache           .                           root.cache
```

# Configuring a Secondary BIND Server

There is some data that has to be stored on the hard disk at a secondary BIND server. A secondary server has its own local copy of the *root.cache* file that lists the root servers. It also has its own dummy primary zone file that takes care of queries for 127.0.0.1. The boot file in Figure 12.15 identifies the directory in which these files are located and lists the names of these files.

In place of file names for the *mgt.com* and *177.207.198 .in-addr-arpa* zones, the secondary boot file provides the IP address of the primary server. The records for these zones will be retrieved from the primary. This server could be made a secondary server for more zones by adding more "secondary" statements to the boot file.

# Server Functions

Before we leave our simple LAN server example, let's review what our servers will do for us:

■ When a LAN user wishes to access an Internet host, our LAN DNS servers will handle the name-to-address query by communicating with an Internet root server and an authoritative name server for the Internet host's domain.

■ Recall that many Internet file transfer and Web servers look up client names and then perform a double-check by translating the name back to an address. Our LAN DNS servers contain the records that will be used to respond to these queries.

■ Our LAN DNS servers will respond to name-to-address translation requests from external users who wish to access our Web server or file transfer server.

We started out defining a pair of NT servers and then switched over to a pair of Unix servers. If you decide to move one or more of your DNS servers, be sure to notify the Registration InterNIC, so that they can update the root servers with the new location of your server. To update a server's name or IP address you must submit a Host template to *HOSTMASTER@INTERNIC.NET.* You can pick up a copy of the template at:

*ftp://rs.internic.net/templates/host-template.txt*

No one will be able to find your new server until you do this chore.

# Configuring DNS for the Internet Environment

We have examined the basics of setting up a Domain Name Server for a small LAN that operates without any firewalls. Now let's take a look at how you would set up Internet Domain Name Servers for a secure network. Suppose that MGT Associates, Incorporated has grown into a very large company with a very large network.

Figure 12.16 shows a sample peripheral LAN configuration for a large organization with Domain Name *mgt.com*. Internal users will exchange mail with Internet users, and will be able to access Internet Web sites via the proxy server.

This network has been set up so that there will be no direct communication between computers in MGT's private internal network and the outside world. Private Domain Name Servers within the boundaries of the internal network will handle queries from internal users about internal systems. Separate, public (sometimes called "fake") databases are used to publicize what you want the outside world to know—and no more.

In the figure, the site's secondary DNS server is located in their ISP's network. This is a common setup. In fact, earlier we saw that both of the public databases for *mcgraw-hill.com* are located in their ISP's network.

The only names and addresses that the outside world needs to know about are the names and addresses of MGT's:

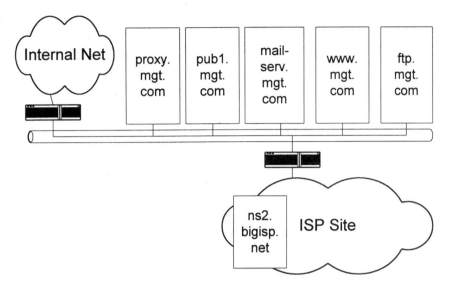

**Figure 12.16**
Sample peripheral
LAN configuration.

- Public DNS servers
- Mail Exchanger
- World Wide Web server
- File Transfer server
- Application Proxy server
- Optionally, an entry might be provided for the external router

Thus, very few entries are needed in the public DNS database. Figures 12.17 and 12.18 show sample zone files.

Note that the Service Provider system, *ns2.bigISP.net*, is listed as a name server near the top of the *mgt.com* zone. However, the address of this system is not listed in this file; the address of *ns2.bigISP.net* must be looked up in the zone file for *bigISP.net*, which is maintained by the Service Provider.

Similarly, the address of *ns2.bigISP.net* is not in the *177.207.198.in-addr.arpa* zone file, because this system is located on a different network.

Figure 12.19 shows another configuration that ABC might use. With the exception of the Mail Exchanger and the Application Proxy server, all of the servers have been located on the ISP's network. Figures 12.20 and 12.21 show the corresponding zone files.

Note that the address-to-name mappings for *www.mgt.com* and *ftp.mgt.com* will be recorded in the ISP's *100.100.204.in-addr.arpa* zone file.

This configuration hides the internal network from the Internet behind MGT's firewall. However, there is one problem that needs to be solved. The

**Figure 12.17**
Public DNS data for
zone *mgt.com*.

```
;
;   Database file mgt.com.dns for mgt.com zone.
;

mgt.com.    IN   SOA   pub1.mgt.com. Admin.mgt.com. (
                      4             ; serial number
                      3600          ; refresh
                      600           ; retry
                      86400         ; expire
                      86400      )  ; minimum TTL


;
;   Zone NS records
;

mgt.com.                        NS      pub1.mgt.com.
mgt.com.                        NS      ns2.bigISP.net.


;
;   Zone records
;
pub1.mgt.com.   5184000 A    198.207.177.100
;
mgt.com.        5184000 MX     10      mail-serv.mgt.com.
mail-serv       5184000 A      198.207.177.101
www             5184000 A      198.207.177.102
ftp             5184000 A      198.207.177.103
proxy           5184000 A      198.207.177.104
```

Mail Exchanger needs to know about external addresses in order to forward
outgoing mail, and it needs to know about internal addresses in order to deliver incoming mail.

The internal DNS servers do not talk to the outside world. If the Mail Exchanger is configured to consult an internal DNS server, it cannot find out
the external information that it needs. If it consults an external DNS, it cannot get the internal information that it needs.

Actually, mail software is so flexible that there are several ways to solve this
problem. But just to show that this problem can be solved quite easily, we'll
present a method that does not depend on any special configuration of your
email software.

Run a DNS server on the peripheral LAN and configure the Mail Exchanger so that it queries only this server. This DNS server also will handle
queries from the proxy server. (In fact, this server could be located at the Mail
Exchanger or the proxy server host.)

Configure this server to act as a secondary for the internal zones. This will
enable it to answer queries about internal systems. However, you wish to

**Figure 12.18**
Public DNS data for
zone *177.207.198.in-
addr.arpa.*

```
;
; Database file 177.207.198.in-addr.arpa.dns for the
; 177.207.198.in-addr.arpa zone.
;

177.207.198.in-addr.arpa.IN SOA   pub1.mgt.com. Admin.mgt.com.
                   3              ; serial number
                   3600           ; refresh
                   600            ; retry
                   86400          ; expire
                   86400        ) ; default TTL


;
;   Zone NS records
;

177.207.198.in-addr.arpa. NS        pub1.mgt.com.
177.207.198.in-addr.arpa. NS        ns2.bigISP.net.


;
;   Zone records
;

100                                 PTR     pub1.mgt.com.
101                                 PTR     mail-serv.mgt.com.
102                                 PTR     www.mgt.com.
103                                 PTR     ftp.mgt.com.
104                                 PTR     proxy.mgt.com.
```

**Figure 12.19**
Moving servers to
the ISP's network.

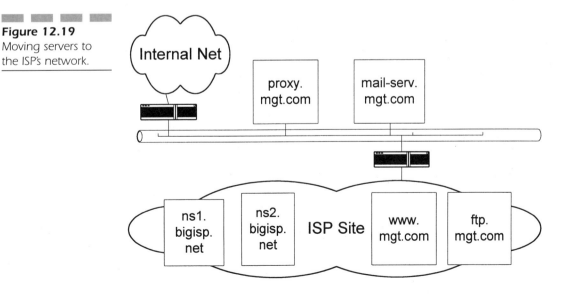

**Figure 12.20**
Public DNS data for
modified zone
*mgt.com.*

```
;
;  Database file mgt.com.dns for mgt.com zone.
;

mgt.com.    IN SOA    ns1.bigISP.net.        Adm.bigISP.net. (
                      4                 ; serial number
                      3600              ; refresh
                      600               ; retry
                      86400             ; expire
                      86400        ) ; default TTL


;
;  Zone NS records
;

mgt.com.                        NS     ns1.bigISP.net.
mgt.com.                        NS     ns2.bigISP.net.


;
;  Zone records
;
mgt.com.           5184000 MX  10 mail-serv.mgt.com.
mail-serv          5184000 A      198.207.177.101
www                5184000 A      204.100.100.50
ftp                5184000 A      204.100.100.51
proxy              5184000 A      198.207.177.104
```

prevent this server from answering queries from anywhere other than the Mail Exchanger and the proxy server. You can take a belt-and-suspenders approach and prevent this in several ways:

- Configure the internal router so that the only DNS traffic allowed to pass in or out of the internal network are the zone transfers for this secondary server.

- Do not list this server in the Internet root directory.

- Configure the external router (and internal router) to screen out all incoming DNS requests, but allow this server to make requests to the Internet and receive responses.

By default, this server would interact with root servers and servers for remote domains in order to get answers to queries from the Mail Exchanger and proxy server. However, if you are concerned about security, you would like to reduce communication with external computers to a minimum.

For extra security, you can configure this server to forward all of its external requests to servers at your ISP. In our example, you could use servers *ns1.bigISP.net* or *ns2.bigISP.net*. Forwarding is configured by adding the

**Figure 12.21**
Public DNS data for
modified zone
*177.207.198.in-
addr.arpa.*

```
;
;  Database file 177.207.198.in-addr.arpa.dns for
;  177.207.198.in-addr.arpa zone.
;     Zone version:   31
;

177.207.198.in-addr.arpa.  IN SOA    ns1.bigISP.net. Adm.bigISP.net
                      3              ; serial number
                      3600           ; refresh
                      600            ; retry
                      86400          ; expire
                      86400      )   ; default TTL


;
;   Zone NS records
;

177.207.198.in-addr.arpa.  NS ns1.bigISP.net.
177.207.198.in-addr.arpa.  NS ns2.bigISP.net.


;
;   Zone records
;

101                             PTR      mail-serv.mgt.com.
104                             PTR      proxy.mgt.com.
```

following statement to the *named.boot* file at this peripheral LAN Domain
Name Server:

```
forwarders      204.100.100.2   204.100.100.3
options         forward-only
```

The "forwarders" statement identifies servers that should be consulted first
for non-local queries. The addresses on the right are the addresses of
*ns1.bigISP.net* and *ns2.bigISP.net*. The forward-only statement says that *only*
these systems may be queried. The peripheral LAN DNS server may not consult the Internet root servers or any other servers.[11]

With this configuration, any external query will be forwarded to one of
these Service Provider DNS servers. That server will obtain the answer to the
query and then send it back. By doing this, we reduce the amount of work that
the DNS server at the mail system needs to do; it will not need to talk to root
servers or other remote servers. We can use our external router to screen out
responses from all sources except for the Service Provider DNS servers.

In the section that follows, we'll look at another configuration that uses
forwarders.

---

[11]This used to be indicated by a "slave" statement in earlier versions of BIND.

# Forwarders

Some sites are not locked up as tightly as the one in Figure 12.16. A site might be using a smart screening firewall that allows internal systems to communicate with Internet servers via TCP sessions. In this case, internal computers need to get name-to-address translations.

One way to do this is to let your internal servers forward queries to your external servers. You set this up just as was done in the previous section, by adding "forwarders" and "options forward-only" statements to the boot files at the internal Domain Name Servers.

```
forwarders      198.207.177.100 204.100.100.3
options         forward-only
```

The filtering routers will have to be configured so that queries can be sent from internal systems to the forwarders and responses from the forwarders are allowed to pass through to the internal DNS servers.

# Why Address-to-Name Zone Files?

In earlier sections we have seen that address-to-name data is entered into zone files that are separate from the files used for name-to-address lookups. For example, there is a zone file that contains address-to-name mappings for the Class C network, 204.151.55. If we want to look up the name of the host that has IP address 204.151.55.44, we would look in this file and discover that this is the address of *www.mcgraw-hill.com*.

But why is it necessary to have this file? The *mcgraw-hill.com* database contains an "A" record that maps *www.mcgraw-hill.com* to 204.151.55.44. Couldn't the Domain Name System use *Whois* to look up 204.151.55, find out that McGraw-Hill owns it, and then use this record to map the address to the corresponding name?

The problem is that McGraw-Hill does not own address 204.15.55. Examine the first few lines of the answer to a *Whois* query for 204.151.55:

```
ANS CO+RE Systems, Inc.  (NETBLK-ANS-C-BLOCK3)
   100 Clearbrook Road
   Elmsford, NY 10523

   Netname: ANS-C-BLOCK3
   Netblock: 204.148.0.0 - 204.151.255.255
   Maintainer: ANS
```

Network 204.151.55 actually is owned by McGraw-Hill's Service Provider, ANS. Like many companies, McGraw-Hill has decided to locate their Web server on their ISP's network. If we do some reverse lookups, we can see how ANS is using this network. We will delete extraneous information from the responses:

```
> nslookup
> set type=ptr
> 204.151.55.44
44.55.151.204.in-addr.arpa    name = www.mcgraw-hill.com

> 204.151.55.45
45.55.151.204.in-addr.arpa    name = www.data.com

> 204.151.55.46
46.55.151.204.in-addr.arpa    name = www.businessweek.com

> 204.151.55.47
47.55.151.204.in-addr.arpa    name = www.osborne.com
```

ANS is operating Web servers for many customers on this network. Thus the hosts on this network have names from many different domains. ANS is responsible for making sure that these servers are running properly, that their communications links are in good shape, and that their DNS records are correct.

This is not a problem because the designers of the Domain Name System were very forward thinking. They decided that the names of systems must not impose any restriction on where the systems are located. To make this work, they required that the address-to-name information be kept in separate zone files. The result is that today, organizations can outsource the operation of Web servers to Service Providers. But the price is that anyone operating a network must maintain a zone file that maps addresses on that network to the corresponding names.

# A Last Word about Zones

Very likely, you will create simple zones similar to the ones in this chapter, which contain name and address information for computers within your organization.

Very large organizations divide this job into more manageable pieces by defining subzones. For example, at Yale, the computer science department operates several large networks in its own right, and there is a separate zone file that contains data for its zone, *cs.yale.edu.*

There are entries in the zone file for the parent zone, *yale.edu*, that identify the name servers for *cs.yale.edu*. This enables lookups to be completed. Let's look at how this works. Suppose that a remote server wants to look up the address of *babyblue.cs.yale.edu*.

- The remote server consults a root server and gets the addresses of name servers for the *yale.edu* zone.
- The remote server sends the query to a *yale.edu* server, and is pointed to the servers for *cs.yale.edu*.
- The remote server sends the query to a *cs.yale.edu* server and gets the answer.

Using subzones requires some special care. A common problem is that a subzone administrator changes the location of a subzone server, but forgets to tell the administrator of the parent zone about the change. The result is that the record in the parent zone file is not changed. As a result, queries are sent to a computer that formerly was used, but which no longer will respond.

# A Tip for Administrators

If you look at the *nslookup* queries in earlier sections, you will see that each computer name was terminated with a period. Formally, a computer name such as *applejack.cs.yale.edu.* is formed by starting at the bottom of the naming tree and writing all the labels that you traverse until you reach the root, which is represented by the period at the end of the name.

When you use *nslookup* to send a query to a name server, *nslookup* will automatically extend an incomplete name by tacking on the local domain name. For example, if I had only entered *applejack*, *nslookup* would have completed it to *applejack.cs.yale.edu.*, a complete domain name. If I enter *www.sun.com*, then *nslookup* will send queries for *www.sun.com.cs.yale.edu* and *www.sun.com.yale.edu* before it sends the correct query. The moral is, don't forget the period.

On a Unix computer, the first choice for the labels used to complete a name is taken from a *domain* statement in *resolv.conf*. If *resolv.conf* does not contain a domain statement, then the assigned hostname is used. For example, if the local host has been assigned the name *plum.cs.yale.edu*, then *cs.yale.edu* would be used for name completions.

On Windows systems, there is a TCP/IP configuration screen that prompts you to enter the domain labels, which are then used for name completion.

# Operations

If you are running a Unix BIND server, be sure to check your online manual entry for *named*. The manual will describe a number of configuration options that may be useful. It also presents a list of signals that you can send to the *named* process. For example, if you have updated some of your zone files, you need to send a "hangup" signal to the DNS process. This will cause the DNS server to reload its zone files and restart.

You need to identify the *named* process by means of its Unix process ID number. You can get the process ID by giving the command:

```
ps -ax | grep named
```

Or, for recent Unix implementations, you can get the *named* process ID from the file */var/run/named.pid.*

Next, use the *kill* command to send a signal to the process:

```
kill -s <signal> <process-id>
```

Some useful signals include:

| | |
|---|---|
| kill -s SIGHUP | Causes the server to read *named.boot*, reload the zone files, and restart. |
| kill -s SIGINT | Dumps the current data base and cached information to the file */var/tmp/named_dump.db*. |
| kill -s SIGIOT | Appends statistics data into */var/tmp/named.stats* if the server was compiled with a suitable option (DSTATS). |

# Security and Performance Issues

## Preventing Unwanted Zone Transfers

By default, the Domain Name System is wide open. Any computer can request a zone transfer from any server. There even is a command built into *nslookup* (*ls -d [domainname]*), which interactively asks for a display of a zone. You may not want it to be that easy for someone to view your whole database. Or you may just want to prevent someone from tying up your server resources by performing unauthorized zone transfers.

A BIND boot file directive can be used to limit zone transfers. For example, the statement below could be added to *ns1.bigISP.net*'s boot file to prevent any system on a network that is external to 204.100.100 from executing a zone transfer from the server.

```
xfernets   204.100.100.0
```

## Fake Responses

One hacker trick was inspired by the fact that some trusting DNS server implementations did not bother to check whether an incoming "response" actually matched a query that they had sent. They would naively store the data in the "response."

The hacker could set up a Web site, copy some pages from a popular site (say, *www.microsoft.com*), and offer virus-laden software as a replacement for the real software download files. The hacker would then pepper a bunch of DNS servers with fake "responses" that contained records that provided the hacker's address instead of the real address of a popular site. These records would be cached.

The solution is to make sure that you run the most up-to-date and reliable version of DNS software at your servers.

## Denial of Service

Hackers always are looking for ways to sneak traffic past your filtering firewall and block your network with junk. An organization that connects to the Internet needs to let DNS responses through its firewall, or else it will be unable to get necessary information about remote sites. (The one exception is a site whose only connection to the Internet is via an application proxy firewall.)

Some hackers try to flood UDP messages that have source port 53 into victim networks. This risk can be reduced by using forwarders and only accessing servers located at your own ISP. The hacker can try to get around this by forging its source address. A good way to prevent this is for other ISPs to check traffic that originates at one of their customers and refuse to carry any datagram that has a bogus source address. A smart firewall can kill any responses that do not match real requests.

## Secure Zones

One or more "secure-zone" records can be added to a BIND zone file. These records are used to restrict access to the information in this zone to specific networks, subnets, and hosts. For example, the records below restrict access to zone information to one specific host address and to systems that belong to network 198.207.177.

```
secure-zone    IN    TXT    "204.100.101.6"
secure-zone    IN    TXT    "198.207.177.0:255.255.255.0"
```

Note that a new record type was not defined. Instead, the secure-zone keyword at the left tells the BIND server to enforce restrictions.

# REFERENCES

The book called *DNS and BIND* by P. Albitz and C. Liu (O'Reilly & Associates, January 1997) provides a good introduction to the Domain Name System.

Currently, the DNS Resources Directory contains the latest news about DNS and pointers to a large amount of reference material. This site is maintained voluntarily by Salamon Andras, who is the technical contact at DNS Services, a South African ISP. It is located at web site:

*http://www.dns.net/dnsrd/*

Primary sources of information include:

- The Registration InterNIC, which provides information and pointers to useful documents at:

*http://rs.internic.net/help/domain/dns.html*

- The Internet Software Consortium, which provides free BIND software and documentation at:

*http://www.isc.org/isc/.*

- The *BIND Operations Guide* or *BOG* (written by Paul Vixie as part of the BIND documentation), which contains detailed and accurate information about Domain Name Servers. It is available at the Internet Software Consortium site.

There are three RFC documents that provide very useful tips for DNS administrators:

RFC 1536 *Common DNS Implementation Errors and Suggested Fixes,* A. Kumar, J. Postel, C. Neuman, P. Danzig, and S. Miller, October 1993.

RFC 1537 *Common DNS Data File Configuration Errors,* P. Beertema, October 1993.

RFC 1713 *Tools for DNS Debugging,* A. Romao, November 1994.


Source RFC documents include:

RFC 1032 *Domain Administrators Guide,* M. Stahl, November 1987.

RFC 1033 *Domain Administrators Operations Guide,* M. Lottor, November 1987.

RFC 1034 *Domain Names—Concepts and Facilities,* P. Mockapetris, November 1987.

RFC 1035 *Domain Names—Implementation and Specification,* P. Mockapetris, November 1987.

RFC 1101 *DNS Encoding of Network Names and Other Types,* P. Mockapetris, April 1989.

RFC 1591 *Domain Name System Structure and Delegation,* J. Postel, March 1994.

RFC 1995 *Incremental Zone Transfer in DNS,* M. Ohta, August 1996.

RFC 1996 *A Mechanism for Prompt Notification of Zone Changes* (DNS NOTIFY), P. Vixie, August 1996.

# 13

# World Wide Web Technology and Performance

The World Wide Web is the application that has turned six-year-olds and great-grandmothers and grandfathers into networking enthusiasts. It also has enabled businesses to set up Intranets that provide easy access to corporate information.

In this chapter, we are going to take a fresh look at hypertext and the hypertext transfer protocol. We will analyze the interactions between Web browsers and servers, introduce a few more performance issues, and discuss some of the benefits of moving from HTTP 1.0 to HTTP 1.1. We will examine a little basic HTML with the object of shedding light on how a Web server works.

Next we'll look at how Web clients are being extended with plug-ins, Java, and ActiveX, and how Web servers are extended with Common Gateway Interface (CGI) programs, modules written to other standard Application Programming Interfaces, Java, and database gateways.

We'll take a close look at the way that forms are created and form data is fed to server programs. Once confidential data starts to flow from a client to a server, there is a need for authentication and confidentiality. We'll walk through the way that the mechanisms introduced in Chapter 10 are used by the Secure Sockets Layer methodology.

In Chapter 14, we will examine two popular Web servers: Microsoft's Internet Information Server for NT, and the free Apache Web server that runs on a multitude of Unix platforms.

# Hypertext Files

Recall that in Chapter 2, we pointed out that Help pages are hypertext files. This means that for each underlined or specially marked phrase or icon, there is some extra hidden "link" information that points to a file that should be displayed when a user clicks on the phrase or icon.

Tim Berners-Lee, the inventor of the World Wide Web, came up with a simple but revolutionary idea:

> *Suppose that when you clicked on an item, the page that you asked for could be retrieved from a different computer?*

Tim Berners-Lee put together all of the ingredients that were needed to make this work.

1. First, he needed a method of writing hypertext pages. He decided to use part of an existing hypertext language, *Standard Generalized Markup Language* (SGML). He customized a subset of this language and called it Hypertext Markup Language (HTML). The original version of HTML was very simple, but was versatile enough to include graphic images and links to local and remote pages.

2. Next, he defined a simple protocol called the *Hypertext Transfer Protocol* (HTTP) that was able to retrieve hypertext pages across a network.

3. Finally, he defined a way to identify the items that we want to retrieve: Uniform Resource Locators or URLs.

When you click on a link on a hypertext page, a hidden URL identifies the location of the information that you have requested. Actually, URLs are not really invisible. When you position your mouse on top of a link, its URL is displayed on the status line at the bottom of your browser screen. A URL:

■ Indicates the protocol that is to be used to get the item.

■ Identifies the host where the item is located.

■ Names the file to be retrieved.

A hypertext URL has the form:

*http://computer-name/filename*

For example,

*http://www.boutell.com/info.html*[1]

You probably never bother to type *http://* at your own desktop; your software inserts this string as the default if no protocol is specified. However, the protocol was included for a good reason. The identification of the protocol to be used has broadened the scope of what can be done with a World Wide Web browser. For example, a link can point to a file transfer server and enable you to copy files using the File Transfer Protocol:

*ftp://www.apple.com/*

Or, a link can allow you to order goods safely by opening a secure session with a commerce server.

*https://www.netscape.com/*

A link can actually point to the name of a computer program. In this case, the program would create a custom-built page for the user.

*http://www.stock-quote.com/cgi-bin/current-quotes.exe*

There are URLs with a somewhat different format. The one below opens up an email application so that you can create a message that will be mailed to the specified destination:

*mailto:webmaster@abc-corp.com*

The flexibility of URLs allows interesting new applications to be added to the Web. The protocols that are needed to support them can be defined as needed.

# Choosing a Web Server

Free or reasonably priced Web server packages are abundantly available. A fairly exhaustive list of servers and a pointer to a server comparison chart can be found in a document at the Boutell site mentioned earlier (*http://www.boutell.com/faq/*).

---

[1]This item points to a document at Thomas Boutell's World Wide Web site, which is a rich source of information about the World Wide Web.

The most popular Unix server is the free Apache server (*http://www. apache.org*). Other well-accepted Unix Web servers include the free NCSA server and the Netscape Communications and Commerce server products. Many sites have purchased the Sun Netra Internet Server, which packages a Netscape server, email service, Domain Name Service, file transfer, and a firewall on a Sun platform. Hewlett-Packard markets a super-secure Web platform called the Virtual Vault. Microsoft's (currently) free Internet Information Server (IIS) is the most popular NT server. An NT server system with IIS provides Web, file transfer, and DNS services along with several other utilities.

# Browsers

The World Wide Web was a wonderful idea, but it really became useful when Marc Andreesen, a student at the University of Illinois, invented the Web client (called a browser) that makes the Web so easy to use. Andreesen's browser became Netscape Navigator. When the world fell in love with the Netscape browser, Microsoft fought back by introducing its Internet Explorer.

Today, the ability to click on a link and zoom to a different document has been built right into desktop applications. Links are embedded in email messages, spreadsheets, and word processing documents.

# Pages

A page is the unit that you obtain when you click on a link to an HTML document.

*An HTML page may be as short as one screen of data, or might occupy hundreds of screens and be as long as a book.*

Often, we access a site by typing in a URL that does not identify a specific file: for example, *http://www.abc-corp.com/*. In this case, we are sent a default page that is called the site's *home page*. Any filename can be used for your home page, but *index.html* and *home.html* are popular choices.

The home page contains links that lead to other pages, and these in turn contain links to further information. We access pages at the site by clicking on selected links.

## ■■■ ■■ **Client/Server Interaction**

The World Wide Web was designed to be a very simple client/server application. A client sends a request to a server. The server processes the request and sends back a result.

You can set up a simple server that allows clients to retrieve static hypertext pages. The top of Figure 13.1 shows what happens when a user selects the URL *http://www.abc-corp.com/products.html*. The client browser connects to *www.abc-corp.com* and asks for the hypertext file *products.html*.

More sophisticated servers enable users to search for data or enter orders for merchandise. In the example illustrated in the lower part of Figure 13.1, first a user fills in a data entry form. Then the client browser connects to *www.abc-corp.com* and asks the server to execute a program named *dbprog*, using the data entered into the form as input. After processing the input, the

**Figure 13.1**
World Wide Web
transactions.

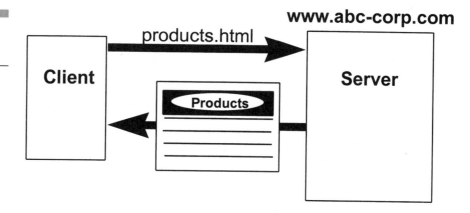

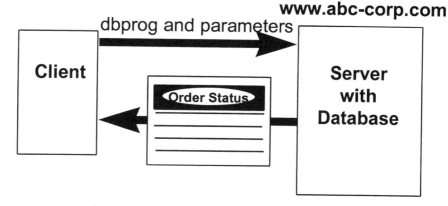

program creates a page that reports search results or transaction status for an order and sends it back to the client.

## Caching Pages

Most browser products save ("cache") pages to hard disk. Also, several pages remain in memory during your browser session. This allows you to go back and review the pages very quickly during your current session. In a later session, if you request a page that has been saved to disk, the browser will contact the server and ask the server to send the page only if it has been modified. If the page has not been modified, the server will send back an "NM" (not modified) response and the cached version will be displayed.

## HTML

We are going to present a little information about the Hypertext Markup Language so that we can explain some important Web server performance issues later. If you can read and understand a little basic HTML, you'll find that it will be easy to understand the impact of specific page designs on server performance and user response time. This information is not intended to be an HTML tutorial. There are lots of tools around that make it easy to create pages in a what-you-see-is-what-you-get fashion. The tools automatically convert your input to HTML.

Web pages are so colorful and attractive that many people are surprised when they learn that the pages are just simple text files that include special instructions called HTML tags. HTML tags:

- Enable a hypertext author to define many document elements, such as its title, section headers, and lists.
- Identify graphics that should be displayed on a page and sounds that should accompany the page.
- Are used to define links to other documents.

At a Unix, Windows 95, NT, or Macintosh system, a text file containing HTML tags usually is called *filename.html, filename.htm,* or *filename.htm3.* At a Windows 3.x PC, an HTML file usually is called *filename.htm.*

**Figure 13.2**
A simple HTML file.

```
<HTML>
<HEAD>
<TITLE>A TEST PAGE</TITLE>
</HEAD>

<BODY>
<H1>WELCOME TO THE OUR WEB SERVER</H1>
<P>
<CENTER><IMG SRC="Rushmore.GIF"
ALT="A picture of Mount Rushmore"></CENTER><BR>

World Wide Web pages can include:
<UL>
<LI>Graphic images
<LI>Sound clips
<LI>Streaming video
<LI>Hypertext links
</UL>
<BR>
Here is a link to
<A HREF="http://www.yahoo.com/"> yahoo. </A>
</BODY>
</HTML>
```

Figure 13.2 displays a printout of a file that contains a very simple HTML page. When displayed, this page contains a picture of Mount Rushmore, a bulleted list, and a link to *www.yahoo.com*.

## HTML Tags

An HTML document includes a HEAD section that contains the title that will be displayed at the top of the user's browser screen. Note that HTML tags are enclosed in angle brackets. Most tags come in start-end pairs.[2] For example:

- The end of the title is </TITLE>. An end tag starts with a "/" symbol.

- The head section can include other background information, such as the author's name and the program that was used to create the file.

- A level 1 heading, marked by <H1>, is displayed in big type. The end of the heading is marked by an </H1> tag.

- The unordered list (from <UL> to </UL>) is displayed using bullets.

---

[2]Tags are not case sensitive. We have used uppercase to make the tags stand out.

Two of the tags are more interesting:

- IMG SRC="Rushmore.GIF" marks the spot where a picture should be inserted. The image is stored in a file called *Rushmore.GIF*. Every image is stored in a separate file.

- ALT="A picture of Mount Rushmore" is the text alternative to the picture. If the user has requested a text-only version of the page, this text will be displayed in a picture frame. After viewing this text, the user may decide that it would be interesting to see what Mount Rushmore looks like, and can retrieve the picture by clicking anywhere inside the picture frame.

- The link to the *Yahoo* search site starts with an "A" tag that is called an "anchor." The format looks a bit odd, but it is clear that it includes the URL of the site to be accessed. The word "yahoo" just before the </A> end tag will appear in color and be underlined. If the user clicks on this word, then a connection will be opened to *www.yahoo.com*.

```
<A HREF="http://www.yahoo.com/">yahoo</A>
```

Other files may be referenced in the hypertext. There might be a file that contains a graphical background to be displayed behind the rest of the page elements. There might be a soundtrack file to be played while the user views the page. There may be files containing Java software that is to be downloaded and executed.

## Getting a Hypertext Page

Let's take a look at the steps that occur when a client retrieves a page. First we'll examine procedures for basic HTTP 1.0. Then we'll see what changes are introduced in HTTP 1.1.

## HTTP 1.0 Client/Server Interaction

After contacting the remote Web server:

- The user's browser begins to receive the HTML file for a page.
- The browser reads the page and starts to display it.
- When the browser comes to a tag that references an image, sound, or Java file, the browser opens a separate TCP connection to the server, requesting the file that holds the item (unless the user has turned off

images, sound, and Java). Older implementations use a separate session for each text file and graphic. Newer implementations still use multiple sessions, but can retrieve more than one item per session.

All of the retrieved components of a page are stored in a cache directory on the client's hard disk. Now we are ready to fill in the technical details. As shown in Figure 13.3:

- The client opens a TCP connection to the server. The well-known port for a World Wide Web server is 80.

- The client sends a command. For example, in Figure 13.3, the client asks for the file *home.html* and announces that its browser supports version 1.0 of the hypertext transfer protocol. The client also announces that it is able to accept hypertext documents. The client could have sent back a longer list of acceptable data types.

```
GET /home.html HTTP/1.0
ACCEPT: text/html
```

- The response consists of a series of headers, a blank line, and the requested item. Tim Berners-Lee borrowed this format from MIME electronic mail, and the Web data types are the same as the MIME data types.

- For version 1.0 of the hypertext transfer protocol, the server closes the connection after the whole document has been transmitted. That

**Figure 13.3**
Detailed Web client/server interactions.

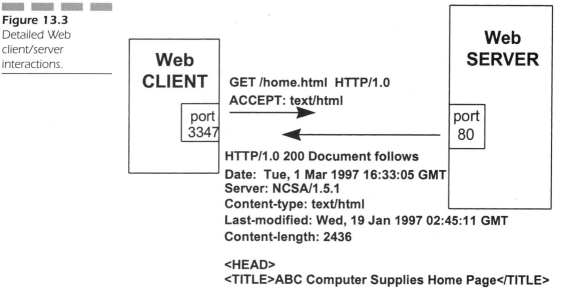

means that while you are reading through a document, you usually are no longer connected to the server.

When you click on a link, your browser can connect to a completely different server without any problem; the old relationship is over.

# Moving to HTTP 1.1

An updated version of the hypertext transfer protocol, HTTP 1.1 was published as a proposed standard in January of 1997. Some vendors had already implemented some of the most needed features. HTTP 1.1 improves performance and the usability of Web interactions for many different types of transactions. Important enhancements include:

- TCP connections can be persistent. By default, TCP sessions will be maintained until either the client or server sends a header parameter that requests a close.

- A client can send a series of requests across a connection without waiting for each reply to arrive. The server will send the responses to these requests in the same order that they were submitted.

- HTTP 1.1 can be used between a client and a proxy Web server when file transfer, news, or mail services are needed. This means that a Web proxy can support proxying for multiple applications.

- Several Web servers can run at the same computer, because the client will send a header that names the server (e.g., *www.abc-corp.com*) that it wants.

HTTP 1.1 adds the flexibility that is needed for multi-step transactions. It removes some of the wasteful overhead caused by opening and closing a large number of sessions to retrieve a page. Typically, more than one session still will be used, however, because tests show that overall response time can be improved by using several sessions.

# More Bad Web Practices

Chapter 2 presented a list of practices that slow your Web server's performance to a crawl, turn off users, and send them away. Now that we know more about the behind-the-scenes technology, let's take another look at Web practices that cause performance and availability problems.

# Effect of a Large Number of Images

We have noted that in HTTP version 1.0, one TCP session is used to download the HTML for a page and each graphic requires an additional separate TCP session. Thus, if a page includes 10 graphics, it will require 11 TCP sessions to obtain the complete page. Even after moving to HTTP 1.1, it is likely that several concurrent sessions will be used to download a page.

Another fact to keep in mind is that while a client and server are setting up a session, new attempts to connect to the server must wait in line—and parking space is limited. For details, review the discussion of listen queues in Chapter 8. The bottom line is that creating pages that trigger a lot of sessions can—and often does—block the entry of other users.

Two recent features can be used to cut back on download sizes.

- Style sheets provide concise information about fonts, colors, and layout.
- Pages can be transmitted in compressed format.

Many of the graphics that appear on Web pages actually present text that has been rendered as a colorful graphic in order to make it more attractive. Style sheets make it possible to describe this information concisely and do away with the need to download a graphic image.

If large images really are needed in order to convey information, the Portable Network Graphic (PNG) standard provides a means to transmit large graphics in an efficient compressed format.

# The TIME_WAIT Problem

Using many short pages or lots of images causes another problem that can be especially troublesome for HTTP 1.0. Recall that the normal pattern for an HTTP 1.0 interaction between a client and a Web server is that:

- The client connects to the server and sends a request.
- The server sends a response and closes.

In Chapter 8, we discovered that the party that closes a TCP session normally goes into a TIME_WAIT state, which freezes all session resources for 4 minutes. Some helpful client browser implementations chop off sessions with resets in order to help the server avoid the TIME_WAIT state. However, if a browser has not been designed to do this, your server will have some of its resources frozen. This is another reason to design your own server so that the number of sessions needed to retrieve information is kept within reasonable bounds. This problem will be eased when client and server software has been rewritten to use the capabilities available with HTTP 1.1. HTTP 1.1 allows the client to initiate the close. TIME_WAITs at a desktop client are not a problem.

In Chapter 8, we also discussed a problem that was the result of clients not sending FINs at the end of a session. Some versions of the Apache Web server that will be described later have had problems stemming from this behavior. Other servers also may share this difficulty. Be sure to check that your server will timeout and recover from these waits.

## Leaving Out ALT Messages

After bringing down the text form of a page, sometimes it appears that one of the images might be useful—but which one is it? Earlier we mentioned that an ALT="Any Text You Like" statement can be included in an IMG statement. This statement will be placed into the picture box. If I am shopping for ski boots, a description like "Women's all-leather ski boots" tells me that this is a link that I should select. Although adding a statement like this is an insignificant effort for the designer, too often it is not done.

## Typographical and HTML Errors

Just about every Web creation tool includes a spelling checker, but a surprisingly large number of professionally built pages contain embarrassing typos. Your company's image sinks from corporate competence to inept beginner very quickly if your pages are peppered with typos.

Errors in HTML, such as forgetting an end tag, cause the ugliest looking pages because mysterious looking HTML instructions will appear on the user's display. For people who are too busy to visually check every page at their site, there are tools (many of which are free) that will do the job automatically.[3]

---

[3]For example, the Spyglass Validator product.

# The Maximum Segment Size

In Chapter 8, we learned that when opening a session, well-behaved client and server TCPs tell one another the size of the biggest segment of data that each is willing to receive. On today's Internet, this size is typically 1460 bytes. Recall that if a client does not announce a maximum segment size, then it is very likely that the server will not send any segment larger than 536 bytes. Some poorly implemented or badly configured server platforms ignore a client's maximum segment size announcement and limit the segment size to 536 bytes when they send segments to remote locations even when they should know better. This slows down performance significantly.

Using larger segments cuts down on the number of messages needed to complete a download and affects response time dramatically. For example, suppose that a short text page along with its World Wide Web headers contained 1420 bytes. If it had to be sent in pieces consisting of at most 536 bytes, then three segments would be needed.[4]

If your site has a page that contains less than 1500 bytes, it may be worth some effort to see if the total response (which includes some hidden headers) can be squeezed into a size that could be delivered in a single 1460-byte TCP segment.

# Flexibility of the WWW Model

The Web started out as a very simple application: connect, get page, close session. But the structure is totally open-ended. Both the client and the server can be extended indefinitely. For example, we mentioned earlier that a URL can be used to start a program at the server.

## Plug-Ins

Initially, clients were extended by adding "plug-in" software to browsers. A "plug-in" is a program that does some special task such as receiving and playing audio or displaying 3-D images. Plug-ins mainly focus on handling special

---

[4]Furthermore, because of the TCP slow start, the server would have to pause and wait for an acknowledgment after sending the first segment.

types of data. Once the plug-in is installed, the user's browser automatically can take appropriate action when the server sends a particular kind of data.

## Java

The Java programming language offers a more powerful way to extend clients by downloading software to the desktop. The user does not perform an installation procedure and execution is not tied to receiving some specific type of data. An instruction to download a Java program file (called an *applet*) is included in a Web page; the code is downloaded and executed immediately. Java can be used to develop programs for any platform. Its portability is a compelling feature.

Java's designers have tried to limit its execution capabilities so that an applet that has been downloaded to a client is walled off from reading or writing local files, executing system commands, or doing anything else that would lead to a security breach. Nonetheless, flaws in the design or implementation of Java are still being discovered and cautious users still block out Internet Java code because of security concerns. Digital signatures and authentication certificates currently are being added to Java implementations. This will help to reassure Internet users that specific Java applets are trustworthy.

Java is a powerful development language and Java programs are used to extend Web servers as well as Web clients. Some servers have been written entirely in Java.

Client and server software can be used to build interactions that follow any pattern that is required. Work is under way to gain agreement on a standardized, coherent, object-oriented framework that will make it easy to extend clients and servers. The ultimate goal is to be able to plug tested and reliable chunks of software (called objects) into an application that needs them. The objects are retrieved from a library of useful software modules.

## ActiveX

Java is not the only way to download code to a browser. Microsoft has added its own ActiveX extensions. ActiveX code is designed to run on Windows desktops, while Java has been implemented across a far wider range of platforms. ActiveX does not have Java's execution limitations, and so is more appropriate to Intranets than the Internet. However, digital signatures and authentication certificates also are being added to ActiveX. These certificates will vouch for the fact that the software came from a reliable source.

# Forms

Web sites can dispense information to users in a convenient way, but the use of forms converts a Web site into a useful business tool. Forms enable a user to enter data and pass it back to the Web server by clicking a *submit* button.

The form itself contains a URL that indicates which of the following should happen when the user clicks *submit*:

- A *mailto* URL causes the form data to be mailed to an indicated recipient.
- An *http* URL identifies the Web server to which the data will be sent, and includes the pathname of a program that will be run when the data arrives.

A form is just a hypertext document that contains form tags. There are tags that identify:

- The start and end of a form
- Text entry fields
- Drop down selection boxes
- Check boxes
- Radio buttons

There are lots of tools that you can use to create forms easily. Even word-processing products like Microsoft WORD and WordPerfect provide easy form builders. We'll look at how a form is built in WORD in order to get an understanding of the major elements.

Building a form in WORD is easy. First, make sure that the Forms toolbar is visible. The forms tool buttons are used to insert text fields, check boxes, drop down boxes, and radio buttons. Figure 13.4 shows the buttons used to create form fields. To start a form, you would choose Form Field from the Insert menu.

**Figure 13.4**
Form field creation buttons.

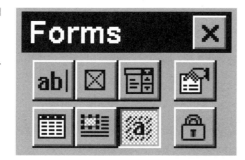

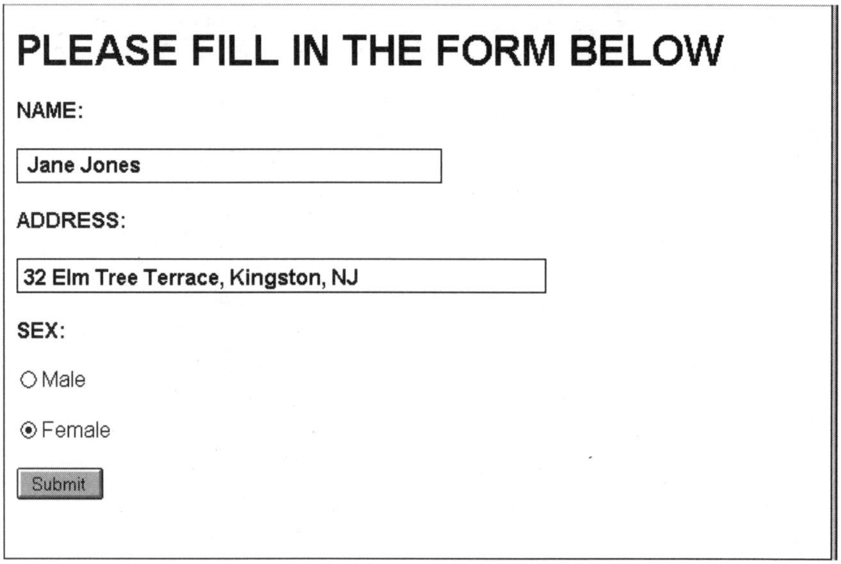

**Figure 13.5**
A typical hypertext
form.

# PLEASE FILL IN THE FORM BELOW

NAME:

| Jane Jones |

ADDRESS:

| 32 Elm Tree Terrace, Kingston, NJ |

SEX:

○ Male

⊙ Female

[ Submit ]

When you click on a form-field button, a pop-up window opens. In the pop-up window, you assign a variable name to the field and describe other attributes, such as the length of a text string item.

After a user has downloaded and filled in this form, the user must click a Submit button to send the information to a Web server for processing. Figure 13.5 shows what a typical form looks like in a browser after data has been entered.

# Submitting a Form

Let's go back one step and examine the information that has to be filled in to create a Submit button. In Figure 13.6, note that we are asked for a URL that identifies what action should be carried out when the form data is submitted.

Recall that there are two choices:

■ The form data can be sent out via email. To do this a *mailto* URL is entered.

■ The form data can be sent to a Web server. In this case, an *http* URL is used.

■ We want our form data to be sent to a Web server, so we have entered the URL:

*http://www.abc.com/cgi-bin/input*

**Figure 13.6**
Configuring form
submission.

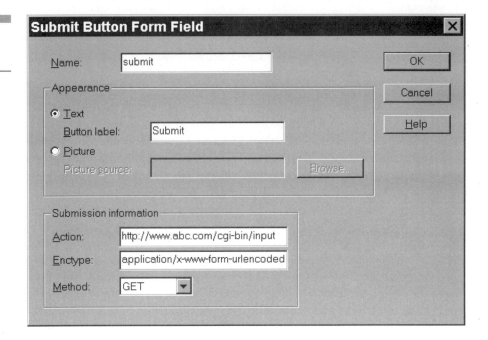

This is the link that is associated with the Submit button. When the user clicks the Submit button, the data that has been input to the form will be bundled up and sent to *www.abc.com*. The filename */cgi-bin/input* identifies the program that will be run at the server. This program will process the data that was entered by the user. The program also will create a page that will be sent back to the user.

## Form Data Format

How is the data sent up to the Web server? A very simple format is used.

*The data is put into a long string that consists of name=value statements separated by ampersand signs.*

Here is how sample input data is formatted before it is sent up to the Web server. Note that spaces within the text field values have been replaced by plus signs:

```
name=Jane+Jones&address=32+Elm+Tree+Terrace,+Kingston,+N.J.&sex=1
```

There actually are two methods of passing this data to the Web server. The default, or GET, method places a question mark at the end of the URL and then adds on the string of data:

```
http://www.abc.com/cgi-bin/input?name=Jane+Jones
&address=24+Elm+Tree+Terrace,+Kingston,+N.J.&sex=1
```

You can watch for this when you use your own browser. After you have filled in a form and clicked Submit, watch the Location window at the top of your screen. You will see these odd looking strings appear.

The second method of sending data up to the Web server is called POST. In this case, the string is not placed at the end of the URL. The string is put into the body of a message that is sent to the Web host.

# Common Gateway Interface

We have seen that when input data arrives at a Web server, we want the Web server to start a program and pass the data to that program. This means that someone must write a program that takes this data and performs a search, queries a database, updates a database, or carries out some other function.

Lots of different vendors build Web servers and applications and so it is important to define standards that provide a way that an application can be added to a Web server. The *Common Gateway Interface* (CGI) was the first industry-standard Web Application Programming Interface (API), and is still the one that is most frequently used. CGI provides some very simple conventions for passing input data to the program identified in a Submit URL.

CGI was designed to be as general as possible. Web site builders want to write programs in many different languages. Languages such as C, C++, or COBOL are in common use. However, many developers like the convenience of a scripting language. Programs written in a scripting language are very easy to write and test. PERL is a very popular scripting language that has been used to write many Web server programs.

CGI was designed so that incoming data could be passed to a program that was created using any programming or scripting language. The CGI solution is:

■ If the GET method is used to send the data to the host, then the string containing the form data is stored into an environment variable at the server that is called QUERY_STRING.

■ If the POST method is used, the data is fed to the program as standard input; that is, as if it had been typed on a command line by a user.

In either case, it is up to the program to interpret and use the information that it has been given. Then the program can generate a suitable page to be sent back to the user.

A programmer should be very careful when writing the software that accepts the input. A favorite hacker trick is to send an extra long input string that overflows the input memory buffer[5] area. A hacker who understands how memory works can plant a pointer to computer instructions at a selected location and cause those instructions to be executed.

This can happen when the programmer has used function calls from the C language or from a scripting language that accepts input of unlimited size. A programmer can prevent this from happening by writing programs that do not use these function calls.

It is common practice to gather all of the Web server programs into one directory. Frequently the name */cgi* or */cgi-bin* is used for this directory. It is important to set the access for this directory to *execute-only*. If the directory is set up so that its files can be *read*, then a hacker can copy down its contents, analyze the programs, and try to figure out what weaknesses they may have. Keep in mind that a hacker can send any type of URL to your site, including a URL that starts one of your programs and feeds it input that may be deadly.

Some newer Web servers that have been built for security wall off all of the add-on programs from the Web server. An administrator configures exactly which programs may be invoked and can add other restrictions depending on user authorizations. The programs run completely separately from the Web server.

## Other Programming Interfaces

Although CGI is popular, many other programming interfaces have been created. We will discuss just a few in this section.

Programmers will have no problem in finding Web development toolkits. Web clients and servers provide such a fertile environment for application development that virtually every software vendor has created Web development tools. The problem that programmers face is in selecting the tools that best meet their needs from an almost overwhelming array of choices.

If an organization has chosen to operate Microsoft's IIS Web server, it is reasonable to make use of Microsoft's *Internet Server Application Programming Interface* (ISAPI). An application that uses Microsoft's ISAPI is com-

---

[5]A buffer is just a memory area that is used to store incoming or outgoing data.

piled as a Windows dynamic link library (DLL) and is loaded into memory when the Microsoft Web server starts up. The program is ready to run as soon as user input arrives, and this gives performance a big boost.

If a Netscape Web server has been selected, Netscape's NSAPI enables a developer to write programs that become integrated into the Netscape Web server, sharing data and other resources with internal server functions or other add-on programs. NSAPI programs can be written to handle form input or to add any general type of capability to a Web server. They are not restricted to handling form data and generating output. HP's Virtual Vault provides a very safe platform for a Netscape Commerce Server. The execution environment for add-on applications is screened off from the Web server.

Applications that are created with Oracle's Web Request Broker API also execute independently of the Web server. The API provides the interfaces that enable the Web Request Broker applications to exchange information with the Web server. As might be expected, an easy-to-use interface to Oracle databases also is provided.

All of the leading database vendors have created packaged "gateway" software that can be used to convert form data into Structured Query Language (SQL) queries and updates. These products usually provide database security and privacy enhancements that are missing from other APIs.

As we mentioned earlier, Java is growing in importance as a way to add functionality to a server. Java can be used with any type of Web server. Many special purpose components are being added to the Java repertoire. For example, the Java Database Connectivity API that makes it easy for a Java program to access any database that supports Microsoft's Open DataBase Connectivity (ODBC) interface. Microsoft's ODBC database interface provides a standard way for programs running on a Microsoft platform to communicate with many different database systems.

As a final note, keep in mind that adding software to your Web server is not something that should be done casually. Poorly written software or programs containing errors can cause performance problems, cause your server to crash, maim valuable data, or open a site to hacker invaders. Software should be reviewed and tested carefully.

# Secure Web Transactions

Secure Web: transactions are being carried out today, and have been possible for quite a long time. There are two competing "standards" specifications for Web security: Secure Sockets Layer (SSL) and Secure HTTP (S-HTTP). SSL is very widely used.

# SSL

SSL was designed by Netscape and was subsequently made available to other vendors. SSL was designed to provide a general layer of security for any application. Although today it primarily is used for secure Web access, it also can be applied to file transfer, *telnet,* news, or other applications. SSL satisfies a number of security needs:

- A server can authenticate itself to the client by means of a certificate.
- The client optionally can authenticate itself to the server by means of a certificate.[6]
- The client and server can communicate privately, using encryption.

When you click on a link that starts a secure transaction, you will see *https* instead of *http* in the window at the top of your screen. If you are using Netscape, you will see that the small broken key in the left-hand corner of the screen has knitted together, and is no longer broken.

By the way, your client will connect to port 443 instead of port 80 when it initiates a secure transaction using SSL. The steps in setting up secure communications (when based on RSA public key encryption) are:

1. The client sends a "hello" announcing what encryption algorithms it supports.
2. The server responds, selecting one encryption method.
3. The server sends its certified public key to the client.
4. The client checks the certification.
5. The client generates a random encryption key.
6. The client sends this key to the server by encrypting it with the server's public key.
7. The server acknowledges that it has received the key and that the negotiation is complete.
8. Data exchanged between the client and server is encrypted using the random key.

If the client also needs to be certified for this transaction, then the client would send its certified public key after step 4.

You can get a peek at this process if you use a Netscape browser. You can check out the server's security credentials by selecting View/Document Info.

---

[6]With SSL-3.

```
Transaction Service has the following structure:

        https://transact.netscape.com:443/ms_dom_bin/txsvr
            Form 1:
                    Action URL:
https://transact.netscape.com:443/ms_dom_bin/txsvr
                    Encoding: application/x-www-form-urlencoded
(default)
                    Method: Post
    . . .
Security:
    This is a secure document that uses a medium-grade
    encryption key suited for U.S. export (RC4-40, 128 bit
    with 40 secret).
            Certificate:
This Certificate belongs to:      This Certificate was issued by:
    transact.netscape.com             RSA Data Security, Inc.
    E-Store Transaction Server        US
    Netscape Communications Corp.
    Mountain View, California, US


                    Serial Number: 02:EE:00:07:04
This Certificate is valid from Mon Nov 11, 1996
to Sat Nov 22, 1997
                    Certificate Fingerprint:
B4:15:9F:24:C1:FC:38:28:8C:6F:21:C2:77:C4:1B:C5
```

Figure 13.7 displays some of the information that appears for a secure page that was retrieved from the Netscape site:

Encryption and decryption are heavy resource users, and doing a lot of it can bog down a server. For this reason, SSL is used only for critical information, such as data entered into order entry forms.

Typically, when setting up a Web server, you would place all documents and programs that will be accessed via SSL (with protocol *https*) into a separate directory or directory tree.

# S-HTTP

S-HTTP was created at Terisa Systems. As its name implies, S-HTTP is not as general as SSL. It was created for use only with *http*. S-HTTP provides for varied kinds of encryption, signature, and key-exchange algorithms.

S-HTTP introduces some new headers and encrypted body parts into HTTP. It is very much in the spirit of a secure MIME specification that also is under consideration by the IETF. Both SSL and S-HTTP have been submitted to the IETF standards process.

# REFERENCES

There are many RFC documents that discuss aspects of World Wide Web technology. Among the most important are:

RFC 2068 *Hypertext Transfer Protocol–HTTP/1.1*, R. Fielding, J. Gettys, J. Mogul, H. Frystyk, and T. Berners-Lee, January 3, 1997.

RFC *2069 An Extension to HTTP: Digest Access Authentication*, J. Franks, P. Hallam-Baker, J. Hostetler, P. A. Luotonen, and E. L. Stewart, January 3, 1997.

At the time of writing, the HTML 3.2 Reference Specification by Dave Raggett was still in working draft form, but it should be available as an RFC by the time of publication of this book.

RFC documents relating to Multipurpose Internet Mail Extensions (MIME) are relevant to the World Wide Web, because they define header formats and document and data types used by both electronic mail and the World Wide Web.

The most up-to-date information on Web client and server technology and emerging standards for hypertext, graphics, and protocols can be found at the World Wide Web Consortium site:

*http://www.w3.org/.*

# 14

# Setting Up a World Wide Web Server

In this chapter we are going to take a closer look at two specific Web servers:

- The server that is packaged with Microsoft's Internet Information Server (which is bundled with Windows NT Server).

- The free Apache Web server used on many Unix platforms.

These two examples should provide a good idea of what is involved in configuring and running a Web server. Although the implementations are greatly different, we will see that there is a core of features that they have in common. The examples are meant to be illustrative. If you need to run a very secure Web server, you should look at some of the customized systems that provide extra protection.

The Microsoft server is extremely easy to install and configure, and is accompanied by extensive documentation and interactive help. The Apache server is complex and the main body of documentation has to be accessed at the Apache Web site, *http://www.apache.org/*. However, the Apache server is flexible and extensible, with so many optional features that you can be pretty sure that it can do whatever you want it to do.

Microsoft's Internet Information Server is integral to Windows NT Server and is immediately ready to run. By default, Microsoft's Internet Information Server, which includes Web services, will start up automatically at boot time. The server is configured and managed via a very simple graphical user interface.

Unix Web servers typically need some customization by a programmer before they can be installed. Often, some source files will need to be edited in order to select features to be included or excluded. Then the code must be compiled. Finally, several configuration files need to be manually edited. The payback for this effort is that the popular Unix Web servers have a lot of flexibility, and performance can be tuned to fit the server's typical transaction mix.

Usually, at a Unix computer several concurrent HTTP server processes are started by entering commands into a script that runs automatically at boot time. This means that several Web server processes already are active when a client request arrives. Any one of these processes can handle the client's request.

# Loading the Pages

It often is most convenient to store all of a server's public pages in one directory and its subdirectories. However, this is not required and the directory structure can be set up in many ways. However, it is important to configure a system and its server software so that no private information is stored in the public area that is open to visitors. Even if no link to a file appears in one of your documents, any file can be accessed if it is in the public area and somebody can guess the file's name.

For example, if some sample programs were included with your Web server product and you did not remove them from the public area, then anyone could connect to your server and run one of these programs, feeding it any input information that they wished.

# Instant Service via Directory Browsing

Suppose that you have some information that you want to make available at a Web server in a hurry, but you do not have time to write hypertext pages. The quickest way to get going is to copy your files into your Web directory, enable "directory browsing," and start the server. You do not need to provide a home page. Figure 14.1 shows the contents of the default "page" that automatically is sent out by a server that has been set up this way. Each name is a hypertext link. Any file can be displayed by clicking on its name. If a user clicks on a subdirectory name (such as *Internet-docs*), then the filenames for that directory will be listed.

The page in Figure 14.1 is not very elegant looking, but the information is available immediately, and that may be the most important consideration.

When you configure your Web server to allow directory browsing, you need to be careful to set up your server so that you strictly control the directories that visitors can access. It is a good idea to check these directories carefully to make sure that you only include files that should be open to the public.

Later, after writing a set of Web pages for your site, you probably will wish to turn this feature off. If you do not, a user still could submit a URL that identifies the name of a public directory instead of a file. The user would be sent a listing showing the contents of the directory, and thus could view and copy any file stored in that directory without any effort.

How could a hacker (or curious user) guess a directory name? Keep in mind that any path to a specific file, such as

*http://www.abc.com/goodstuff/programs/dataentry*

reveals information about your directory structure. You can be sure that some hacker will try to open a URL such as

*http://www.abc.com/goodstuff/*

just to see if anything interesting turns up. Hence, it is important to:

- Use Web servers that let you strictly limit clients to a list of permitted directories.
- Use Web servers that let you prohibit directory browsing if you want to.
- Not permit directory browsing unless you have nothing to hide.

**Figure 14.1**
Default page
displaying directory
contents.

| | | | |
|---|---|---|---|
| 12/11/96 | 6:29 | AM | 22462 countrycodes.txt |
| 11/16/96 | 6:06 | AM | 35785 history.txt |
| 1/4/97 | 3:10 | PM | \<dir\> Internet-docs |
| 12/7/96 | 1:53 | PM | 1062 info.txt |
| 1/9/97 | 10:10 | AM | 129180 rfc1034.txt |
| 12/11/96 | 6:28 | AM | 706 root-servers.txt |
| 12/27/96 | 9:11 | AM | \<dir\> samples |

# Running Multiple Web Servers on One Computer

As we have noted, some organizations register more than one name. For example, McGraw-Hill has registered both *mcgraw-hill.com* and *mgh.com* at a single computer. For HTTP 1.0, this is done by using a trick that happens to work. You assign more than one IP address to the system's network interface and associate a different computer name with each address.

Many Internet Service Providers have assigned several IP addresses to a system's network interface so that one of their computers can be used to host Web servers for several customers. Then new name and address entries are added to the Domain Name System. Multiple addresses will not be needed after HTTP 1.1 is widely adopted, because each client will send a header that names the server that it wants.

Figure 14.2 shows clients connecting to *www.abc.com* and *www.def.com,* which are located at the same computer. The Web server at the host has been configured to associate a directory subtree and default home page with separate IP addresses (that have been assigned to the same interface).

**Figure 14.2**
Two Web servers located at the same host.

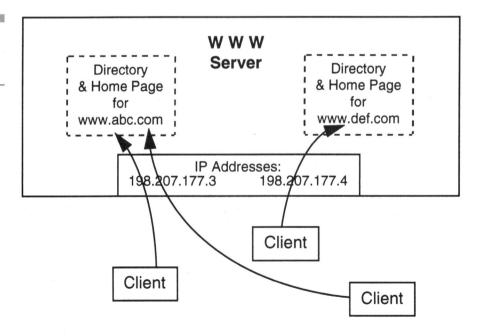

# Microsoft's Internet Information Server

The best time to prepare an NT Server system to act as a Web server is when you are installing the operating system. Two actions put you on the right track:

- Choose NTFS as your file system. It is the only choice that offers file security.

- Make sure that you check the box that asks whether Microsoft's Internet Information Server (IIS) should be installed. If you do, IIS will be installed, set up, and started automatically.

# The IIS Servers

The Internet Information Server actually includes three servers: World Wide Web, file transfer, and gopher. (Gopher is a service that has been made obsolete by the World Wide Web, but a few sites continue to use it.) By default, the installation automatically creates a server root directory called *InetPub,* which contains a directory for each application: *wwwroot, ftproot,* and *gophroot.*

By default, your servers are automatically started at boot time. However, if IIS was not installed with the operating system and you now want to add and use it, first be sure to check the system to make sure that no one has installed some other WWW, FTP, or gopher software. To do this, choose **Services** on the **NT Control Panel** and scroll through the list. If one of these servers is listed, select it and click the **Startup** button. You should then choose the **Disabled** radio button to disable the unwanted application. You can then choose the corresponding Microsoft IIS service, click on **Startup**, and choose the **Automatic** button.

You must disable the old server because otherwise it will grab the port for the application the next time that your system boots. For example, if there is a pre-existing Web server on the system, it will grab port 80 the next time that you boot your system and this will prevent your Microsoft IIS Web server from communicating with any clients. Following the principle of keeping minimal software on a public server, you should remove any unused application software from the system.[1]

---

[1] To remove software, use **Add/Remove Programs** on the **Control Panel**.

Once your IIS server has been installed and started, users will be able to connect to it immediately. If you want to defer access until you have finished building your site, the best strategy is to detach the server computer from the network during the construction and test period.

Few people are interested in operating or using gopher servers today. If you are not going to support gopher, use Services on the Control Panel to disable it. You will not be able to remove the gopher software because all IIS services are integrated into a single component.

# Creating the Web Site

Before you start, take time to do some planning. A site that has an organized structure will be easier to set up and to maintain. Give a lot of thought to maintenance. This is a big job. You will need to set policies that control what can be published and how long it stays published. You need personnel to review and test any software that needs to be built. You need to try out and select tools that display your Web map, check your links, and track your inventory.

## A Simple Public Site

Some sites are very simple. They just provide fairly static pages that users can read. The easiest way to create this type of Web site is to create a convenient directory subtree structure under *wwwroot* and then put your information into the *wwwroot* directory and its subdirectories. All document directories should be configured as read-only and program directories should be execute-only. Remember to remove all of the unused samples that are shipped with the IIS product, especially the sample scripts and programs.

Even a simple document server can get out of control if it gets big enough. Some sites use a database of documents instead of a simple directory tree.

## Interactive Public Site

Sites that will present forms to users so that they can interact with visitors are more interesting, but also are a lot more work to set up. Recall that the form data will be returned along with a URL that identifies the program to be executed. The program creates the page that the Web server will send back to the user.

Remember that it is prudent to isolate all of your Web programs into a single directory subtree, and restrict its access to execute-only. This means that the programs can be run on behalf of users but nobody will be able to copy them.

## Other Directories

You can define links to files in other directories outside of the *InetPub* tree. In fact, the directories can even be located at other NT servers. Directories located outside of the *InetPub* tree are called virtual directories. Keep in mind that *any* file located in a virtual directory potentially could be retrieved. You will have to keep an eye on the contents of the virtual directories and control access to them carefully. If you have a compelling reason to use virtual directories, you will need to add their names to the list of Web server directories. Microsoft Internet Server Manager provides a screen that enables you to do this.

You must assign each virtual directory an alias name. A URL will point to an external file using the alias. To users, each virtual directory will appear to be a subdirectory of the Web root.

For example, suppose that you operate site *www.abc-corp.com* and have written new HTML files in a directory named:

*C:\additional\new_web_stuff\sports.v12.*

If you assign the alias name */sports* to that directory, a user can retrieve a file called *swimming.htm* with the URL:

*http://www.abc-corp.com/sports/swimming.htm*

## Restricted Site

There are a number of Internet sites that serve only paid subscribers. There are two ways to do this. One method is to use a password-based challenge handshake protocol for authentication, similar to that described in Chapter 10. There is a proposed Internet standard that describes how to implement this in clients and servers so that products created by different vendors can use this method. Another method is to use SSL and certificates for client identification. Client certification is not widely used today but it will become commonplace in the future.

The Microsoft IIS server can be configured to support several categories of users, each with different access privileges. At the time of writing, access control could be supported by either:

■ Sending a username and password in the clear

■ Using a proprietary Microsoft challenge handshake

## Housecleaning

It is a good idea to check periodically to make sure that there is nothing in your public directories that you do not want the world to see and that information that should only be available to authorized users is appropriately protected. When you install a new version of a program, remember to remove the old one.

## Testing

Recall that you can test your Web service using a browser that runs at the same computer by opening a connection to *localhost* or 127.0.0.1.[2] This makes it easy for you to view the pages at your site and check the flow of information.

If you have any programs at your site, they need to be examined and tested very carefully. The programs may be intended to run with data that users have entered into forms. However, when a URL containing a program name and a block of input parameters arrives at a Web site, you have no way of knowing whether it came from a data entry form or from a hacker hoping to make trouble. Hackers have used naively written programs to do anything from crashing a server to breaking into the computer.

# NT Web Server Administration

There are several administrative tools that an NT Web administrator needs to know about. These tools are listed in Table 14.1. We'll provide information about some of these services in the sections that follow.

---

[2]See Chapter 7.

**TABLE 14.1**

*NT Administrative Tools*

| Tool | Description |
|------|-------------|
| Microsoft Internet Server Manager | Configure service parameters, define the directories to be used, and start and stop WWW, FTP, and gopher. |
| Services Application | View and change the status of all system services, and configure or disable automatic startup. |
| Task Manager | View the list of active processes, view process and system performance. |
| Performance Monitor | Measure Web server activities at whatever level of detail is desired. |
| Netstat | Display active sessions and provide network statistics. (Run from a command screen.) |
| Network Monitor | Capture selected traffic to and from the server, track statistics, and display frame-by-frame protocol analyses. |

# Microsoft Internet Server Manager

The Microsoft Internet Server Manager provides a central point of control for one or more IIS servers. From the manager screens, you can pause, stop, or start services. You can select services and view or change their operating parameters. If you need to protect traffic via encryption, there is a menu choice that enables you to define a public/private key pair.

## NT Computer Name

NT Servers are identified by name, but the name that appears is the Microsoft name, not an Internet Domain Name System name. It is easy to get confused about the difference between an NT server name and a Domain Name System name, so we will digress slightly to explain this point.

During NT Server system installation, the administrator has to enter a one- to fifteen-character name that identifies the system for the purposes of Microsoft networking protocols. It is up to the system administrator to decide whether this name should be reused as the first label of the computer's Domain Name.

The names can be completely different. You might decide to use a Microsoft name of NTSERVER, a DNS name of *server1.abc-corp.com,* and a DNS alias name of *www.abc-corp.com.*

## NT IIS Web Server Access Controls

As we have noted, often a very simple configuration is used for a Web server. Any user may connect from anywhere. Public information stored in a default directory is available to everyone. Sometimes, you want to screen out a few troublemakers. Or, you might actually wish to permit only a selected set of specific sites to access your server. You can list specific addresses, subnets, or networks that are granted or denied access to the server. But keep in mind that screening by IP address is a very feeble form of security.

Private members-only services can be supported by adding usernames and group definitions to the system and using standard NT administration tools to define the privileges for these users and groups. As mentioned earlier, an emerging standard for a challenge handshake will support the ability to authorize special services. Some sites do this today by accepting usernames and passwords that are sent across the Internet in the clear, which is not a good idea.

## WWW Parameters

You use Microsoft Server Manager[3] to view and set server parameters. If you double-click on a World Wide Web server entry, you can access a number of parameters. Several of the configuration parameters are described in Table 14.2.

## Setting Up SSL

It is not difficult to set up SSL at an NT IIS Web server. However, you need to get certified by a responsible authority so that users can be sure that your site is trustworthy. There will be a fee for the certification service. Let's look at the steps that you need to follow in order to prepare for SSL.

1. Create directories to hold pages that must be sent in encrypted form and programs that will operate on private form data.

2. In Internet Service Manager, double click on the Web server and make sure that the secure directories are in its directory list.

---

[3]At the current time, Microsoft Internet Server Manager is initiated via Start/Programs/Microsoft Internet Server (Common)/Internet Service Manager.

**TABLE 14.2**

IIS Configuration
Parameters

| Parameter | Description |
|-----------|-------------|
| TCP Port | The default is 80. |
| Connection Timeout | Timeout before dropping an inactive user. |
| Maximum Connections | Maximum number of concurrent connections. |
| Types of authentication used | Currently anonymous (public) access, clear text username and password, or Windows challenge/response. |
| Anonymous username and password | Authentication information for the account that defines the privileges given to anonymous users. |
| WWW Directories | Server directories. If multiple servers (*e.g., www.abc.com, www.efg.com*) will run at the same host, the IP address that corresponds to the appropriate server is associated with each directory. One directory will be identified as the home directory for this IP address. |
| Default page for each directory | The name of the file containing the default html page that will be displayed when a user connects to a directory without specifying a file name.  For the home directory, it is the default home page displayed when a user connects to that Web server (e.g., with *http://www.abc.com/*). |
| Directory browsing | Recall that if directory browsing is enabled, a user can send a URL that names a directory, can view the list of files in the directory, and can request any of the files. |
| Directory or database to hold log messages | Log entries can be recorded at the local host or can be sent to a different computer. |
| Limit on the maximum outgoing traffic from this server | For example, if your Web server is located at your own site, you can make sure that the server shares your connection to the Internet with electronic mail traffic and with sessions between internal clients and Internet servers.<br><br>When outgoing traffic from the server gets close to the threshold, the server will delay its responses. If the bandwidth is exceeded, then the server will refuse requests to read files until the bandwidth drops to the threshold. |

3. Generate a public/private key pair with the Key Manager tool.

4. Get your public key certified.

5. Install your certificate.

6. Go back to the Internet Server Manager's Web directory list, select a directory that needs to be protected with SSL, click Edit Properties, and check the box that says Require secure SSL channel.

We'll take a closer look at steps 3, 4, and 5 in order to show that the procedures are simple and make sense. In order to establish a trusted Web site, you need to generate a public/private key pair, and you need to get certified.

Why is certification needed? You are going to use SSL so that confidential client information will be hidden from the rest of the world. But they are sending this information to *you*. How do they know that you are not a hacker or a thief? The Internet is a wide open place and there is nothing that prevents dishonest people from setting up Web sites.

Certifying a Web server is the process of getting a trusted third party to vouch for your identity. Think of the process as similar to becoming a credit card merchant. Before you are authorized to accept credit card payments, someone checks up on your identity, address, and credit worthiness. When the process is complete, you will be permitted to display a certificate showing a VISA or MasterCard logo on the front window of your premises. Similarly, a certified Web site will display a certified logo that identifies the Certification Authority that has validated your business.

## Generating Keys and a Certificate Signing Request

It turns out to be very easy to generate a public/private key pair for a Web server. For NT IIS, there is a **Key Manager** selection on the Internet Server Manager's **Tools** menu. The **Key Manager** creates your key pair and prompts you for information that it uses to automatically create a certificate signing request (CSR) file. This file contains some of the information that you need to provide when you apply for certification.

Figure 14.3 shows the format of Microsoft's certificate signing request file. We have added italicized descriptions of the information to be entered into some of the fields. Only very basic information is required.

The certification procedures will vary depending on whom you choose to be your Certification Authority. Let's walk through Verisign's procedures. The Verisign Certification Authority has set up pages that make it easy to request certification for many types of Web servers, including Microsoft's.

The steps are:

■ Email the request to *microsoft-request-id@verisign.com*.[4]

■ Verisign will send back an encoded version of your certificate signing request via electronic mail.

---

[4]Verisign has set up mailboxes for other server products, for example, *netscape-cert@verisign.com*.

**Figure 14.3**

A certificate request.

```
Webmaster: (email address)
Phone: (phone number)
Server: Microsoft Key Manager for IIS 2.0

Common-name: (Domain name of Web server, e.g. www.abc.com)
Organization Unit: (Specific division, e.g., ABC Machine Tool
[optional])
Organization: (Legal name of the organization, e.g., ABC
Equipment Corp.)
Locality: (Town or City. Optional for multilocation
organizations)
State: (Operating location: Do not abbreviate)
Country: (2-character ISO format country code)

-----BEGIN NEW CERTIFICATE REQUEST-----
(Your public key belongs here.)
    . . . . . . . . . . . . . . . . . . . . . . . . . .
    . . . . . . . . . . . . . . . . . . . . . . . . . .
    . . . . . . . . . . . . . . . . . . . . . . . . . .
    . . . . . . . . . . . . . . . . . . . . . . . . . .
    . . . . . . . . . . . . . . . . . . . . . . . . . .
    . . . . . . . . . . . . . . . . . . . . . . . . . .
    . . . . . . . . . . . . . . . . . . . . . . . . . .
    . . . . . . . . . . . . . . . . . . . . . . . . . .
    . . . . . . . . . . . . . . . . . . . . . . . . . .
-----END NEW CERTIFICATE REQUEST-----
```

- Connect your Web browser to Verisign (*www.verisign.com*), select their Digital ID Center, and provide the requested information.

  You will be prompted for information such as:

- The plaintext version of your certificate signing request along with the encoded version that they sent you via electronic mail.

- Information about your technical, administrative, and billing contacts.

- A Dun & Bradstreet DUNS number or equivalent information, such as your articles of incorporation.

- A challenge phrase that will protect future actions.

The Certification Authority will check that your company information matches the registration information for the server's domain name, and it will validate the DUNS number (or equivalent) information.

After checking your identity, Verisign will create a certificate for your public key, add its own certificate and a digital signature, and email it back to your technical point of contact. The certificate should then be saved into a

**Figure 14.4.**
A public key
certificate.

```
-----BEGIN CERTIFICATE-----

JUYTRDo45667XXCwLQMJSoZILvoNVQECSQAwcSETMRkOAMUTBhMuVrM
LkoAnBdNVBAoTF1JTQSBEYXRhIFN1Y3VyaXR5LCBJbmMuMRwwGgYDVQ
QRExNQZXJWmnbvIENlcnRpZmljYXRlMSQwIgYDVQQDExtPcGVuIE1hc
bghytwasXN0IFNlcnZlciAxMTAwHhcNOTUwNzE5MjAyNzMwWhcNOTYw
poY678R5dTEwWjBzMQswCQYDVQQGEwJVUzEgMB4GA1UEChMXU1NBIER
hhydgU2VjdXJpdHksIElKqr4DfDAaBgNVBAsTE1BlcnNvbmEgQ2VydG
bghiuHHkjlJDAiBgNVBAMTG09wZW4gTWFya2V0IFRlc3QgU2VydmVyI
DExMDBcMA0GCSqGSIb3DQEBAQUAA0sAMEgCQQDU/7lrgR6vkVNX40BA
q1poGdSmGkD1iN3sEPfSTGxNJXY58XH3JoZ4nrF7mIfvpghNiltaYim
vhbBPNqYe4yLPAgMBAAEwDQYJKoZIhvcNAQECBQADQQBqyCpws9EaAj
KKAefuNP+z+8NY8khckgyHN2LLpfhv+iP8m+bF66HNDUlFz8ZrVOu3W
Pouytr5dStrskNKXX3a

------END CERTIFICATE-----
```

file. The encoded certificate will look similar to the block of data shown in Figure 14.4.

To install your certificate, you go back to **Key Manager**, select the key, and choose **Key/Install Key Certificate**. You will be prompted for the name of the file that contains the certificate. Once your key and certificate are installed, back them up to a floppy disk or tape using the **Export Key** command. If anything goes wrong, you can import them later.

After your certified keys have been set up, a URL that arrives at your server and requests a document or program in an SSL directory must start with *https*. If it does not, the request will fail.

# Security Tips

## Security Checklist

There are several items that need to be checked before you open up your site for business. You should:

■ Unbind unnecessary services from the network interface.

■ Make sure that the only user accounts on the system are those created for the system administrators.

- Disable the NT built-in account with username *guest*. Its purpose is to allow someone to access the NT server temporarily without requiring a new account to be set up. That person would have enough privilege to try to poke around, looking for directories or files that have not been protected.

It is easiest to keep a Web server secure when its end-user access is restricted to anonymous users who are limited to specific directories. Allowing remote access by other user accounts opens the system to attempts to break in by guessing usernames and passwords.

## Auditing

Human error and operating system holes make it impossible to guarantee that any server is 100% buttoned-up. The NTFS file system supports auditing of attempts to access files and folders. Audit reports are a very valuable resource that can:

- Detect when break-ins are being attempted.
- Assist in damage repair if a break-in has succeeded.

## Internet Username

When anonymous Internet users access your public Web, file transfer, or gopher server, they will have the privileges that have been assigned to a fictitious username. By default, the username IUSR_*computername* is automatically created for this purpose and used for all three services.[5] The *computername* is the Microsoft name (e.g., NTSERVER, as above). It is a good idea to include the Microsoft system name in the dummy username to make sure that the user identifier is unique throughout your Microsoft networking environment.

Suppose that you wanted file transfer users to have different privileges than Web users. The best way to do this is to create a new account that is a copy of the *IUSR_computername* account. The new account will be created with an appropriate set of security restrictions, and then you can make whatever changes are needed to assign the new username to the appropriate service.

---

[5]A random password is assigned to this account.

# NT Server Performance Monitor

NT includes a Performance Monitor that can track information about the operating system and its supported services. Currently, you open the performance monitor by selecting:

Start/Programs/Administrative Tools (Common)/Performance Monitor

To track Web activities, select Edit/Add To Chart, scroll through the choices in the Object box to HTTP Service, and then choose the Web counters that you wish to activate. Currently, over 25 counters are defined, including items such as:

- Bytes Received/sec
- Bytes Sent/sec
- Bytes Total/sec
- CGI requests
- Connections/sec
- Current Anonymous Users
- Current connections
- Files Sent
- Get requests
- Maximum Anonymous Users
- Not Found errors

Detailed reports on the internal workings of the server can be obtained by choosing Internet Information Server Global in the Object box. Current selections include items such as:

- Cache Hits
- Cache Used (in shared memory)
- Cached File Handles
- Measured Asynch I/O Bandwidth Usage (averaged per minute)
- Current Blocked Async I/O Requests (because of bandwidth throttling)

# NT Server Network Monitor

NT Server includes a Network Monitor that enables you to get an overview of traffic to and from the server and capture selected streams of frames. This tool assists on performance assessment, network troubleshooting, and security troubleshooting.

# Apache Server

The Apache server was created and is maintained by a dedicated group of volunteers. Their purpose is to provide a public domain implementation that conforms to HTTP standards. The current version of Apache supports HTTP 1.1. The Apache server is a strong performer with lots of flexibility, but as you might expect with free Unix software, the Web administrator will have to understand the Unix environment, should be a competent programmer, and must be willing to invest some effort up front.

Binary executable code is available for some platforms, but many administrators prefer to start with source code. When starting from source, the administrator initiates the installation process by editing a configuration file in order to select:

- Operational rules to be used.
- Optional modules to be included in the server.

There is an API that you can use to create your own modules. If you have done so, you can include your customized software modules at this point.

Next you would run a script that creates an appropriate *makefile,* and finally, you would type the Unix *make* command that compiles and links the application.

Configuration now starts in earnest, and at the time of writing, there is no graphical user interface to prop up this process. However, sample versions of the text files that need to be edited in order to configure the server (*srm.conf, access.conf,* and *httpd.conf*) contain frequently used defaults and are well commented. Dozens of directives that customize the Apache server's operations can be written into these files. Directives have a simple format and many examples are provided. For example, the following statement identifies the top of the directory tree that contains the server files and documents:

```
DocumentRoot /home/httpd/html
```

**TABLE 14.3**

Apache
Configuration
Parameters

| Parameter | Description |
|---|---|
| TCP port to be used | Default 80. |
| Timeout | For dropping an inactive user. |
| Minimum and maximum for the server pool size | Typically, several server processes run at the same time. Apache automatically adapts to its current load, keeping extra processes running to handle traffic spikes. These minimum and maximum values are used to periodically readjust the number of processes. |
| Maximum clients | Maximum number of concurrent clients. |
| The user and group name under which the server will run | This will define the privileges given to anonymous users. |
| The document root for the server | If several servers run at the system, separate directories are identified for each server. |
| Default page for the directory | This file will be displayed when a user connects to the directory. |
| The directory where the server's config, error, and log files are stored | If several servers run at the system, separate directories are identified for each server. |
| For each directory, allow and deny access controls | This can include specification of whether *cgi* functions stored in the directory may be executed. |
| Further access limits for directory access | These can be defined by another file called *.htaccess* that is stored in the directory. |
| Usernames and passwords | Authenticated access can be established by creating a username and password file that is independent of the system's usernames and passwords. Separate files also are used to set directory access limits for these users. However, only basic authentication (username and password sent in the clear) is supported at the time of writing. It is expected that a challenge handshake method will be implemented. |

Table 14.3 lists some of the configuration parameters that can be set in these files. Several of these parameters coincide with parameters defined for the NT server. However, the list in the table is just a sampling. Many more optional parameters can be used with the Apache server.

Apache has many extra capabilities. To name just a few, it can be configured to:

- Create pages that vary according to the type of client browser.
- Execute CGI scripts under a different (e.g., very low privilege) user-name.
- Set up a log for CGI debugging.
- Establish resource limits for CGI scripts.
- Associate a script with a file type. When a request for a file of that type is sent to the server, the script is executed. For example, a script could be used to customize some pages to include a news flash of the day.
- Send customized error responses to clients.

This server is clearly far more complex than the NT server, but many Unix administrators have selected the Apache server because of its power, and because they know that if a server bug or security problem is discovered, they will be able to repair it immediately since they have control of the source code.

# REFERENCES

*Microsoft NT Server Resource Kit,* Microsoft Press, 1996.

Apache documentation is at *http://www.apache.org/.*

# APPENDIX A

## TRANSLATING BETWEEN BINARY AND DECIMAL

To translate a binary number to decimal, you need to multiply each binary digit by an appropriate power of 2. Below, we show the multipliers that are used to convert 10010111 to decimal.

```
  1    0    0    1    0    1    1    1
128   64   32   16    8    4    2    1
```

Thus, 10010111 is $(128)(1) + (64)(0) + (32)(0) + (16)(1) + (8)(0) + (4)(1) + (2)(1) + (1)(1)$, which equals $128 + 16 + 4 + 2 + 1 = 151$. The table that follows displays all 8-bit translations.

Binary to Decimal Translation Table

| Binary | Decimal | Binary | Decimal | Binary | Decimal | Binary | Decimal |
|--------|---------|--------|---------|--------|---------|--------|---------|
| 00000000 | 0 | 00100000 | 32 | 01000000 | 64 | 01100000 | 96 |
| 00000001 | 1 | 00100001 | 33 | 01000001 | 65 | 01100001 | 97 |
| 00000010 | 2 | 00100010 | 34 | 01000010 | 66 | 01100010 | 98 |
| 00000011 | 3 | 00100011 | 35 | 01000011 | 67 | 01100011 | 99 |
| 00000100 | 4 | 00100100 | 36 | 01000100 | 68 | 01100100 | 100 |
| 00000101 | 5 | 00100101 | 37 | 01000101 | 69 | 01100101 | 101 |
| 00000110 | 6 | 00100110 | 38 | 01000110 | 70 | 01100110 | 102 |
| 00000111 | 7 | 00100111 | 39 | 01000111 | 71 | 01100111 | 103 |
| 00001000 | 8 | 00101000 | 40 | 01001000 | 72 | 01101000 | 104 |
| 00001001 | 9 | 00101001 | 41 | 01001001 | 73 | 01101001 | 105 |
| 00001010 | 10 | 00101010 | 42 | 01001010 | 74 | 01101010 | 106 |
| 00001011 | 11 | 00101011 | 43 | 01001011 | 75 | 01101011 | 107 |
| 00001100 | 12 | 00101100 | 44 | 01001100 | 76 | 01101100 | 108 |
| 00001101 | 13 | 00101101 | 45 | 01001101 | 77 | 01101101 | 109 |
| 00001110 | 14 | 00101110 | 46 | 01001110 | 78 | 01101110 | 110 |
| 00001111 | 15 | 00101111 | 47 | 01001111 | 79 | 01101111 | 111 |
| 00010000 | 16 | 00110000 | 48 | 01010000 | 80 | 01110000 | 112 |
| 00010001 | 17 | 00110001 | 49 | 01010001 | 81 | 01110001 | 113 |
| 00010010 | 18 | 00110010 | 50 | 01010010 | 82 | 01110010 | 114 |
| 00010011 | 19 | 00110011 | 51 | 01010011 | 83 | 01110011 | 115 |
| 00010100 | 20 | 00110100 | 52 | 01010100 | 84 | 01110100 | 116 |
| 00010101 | 21 | 00110101 | 53 | 01010101 | 85 | 01110101 | 117 |
| 00010110 | 22 | 00110110 | 54 | 01010110 | 86 | 01110110 | 118 |
| 00010111 | 23 | 00110111 | 55 | 01010111 | 87 | 01110111 | 119 |
| 00011000 | 24 | 00111000 | 56 | 01011000 | 88 | 01111000 | 120 |
| 00011001 | 25 | 00111001 | 57 | 01011001 | 89 | 01111001 | 121 |
| 00011010 | 26 | 00111010 | 58 | 01011010 | 90 | 01111010 | 122 |
| 00011011 | 27 | 00111011 | 59 | 01011011 | 91 | 01111011 | 123 |
| 00011100 | 28 | 00111100 | 60 | 01011100 | 92 | 01111100 | 124 |
| 00011101 | 29 | 00111101 | 61 | 01011101 | 93 | 01111101 | 125 |
| 00011110 | 30 | 00111110 | 62 | 01011110 | 94 | 01111110 | 126 |
| 00011111 | 31 | 00111111 | 63 | 01011111 | 95 | 01111111 | 127 |

Binary to Decimal Translation Table (Continued)

| Binary | Decimal | Binary | Decimal | Binary | Decimal | Binary | Decimal |
|---|---|---|---|---|---|---|---|
| 10000000 | 128 | 10100000 | 160 | 11000000 | 192 | 11100000 | 224 |
| 10000001 | 129 | 10100001 | 161 | 11000001 | 193 | 11100001 | 225 |
| 10000010 | 130 | 10100010 | 162 | 11000010 | 194 | 11100010 | 226 |
| 10000011 | 131 | 10100011 | 163 | 11000011 | 195 | 11100011 | 227 |
| 10000100 | 132 | 10100100 | 164 | 11000100 | 196 | 11100100 | 228 |
| 10000101 | 133 | 10100101 | 165 | 11000101 | 197 | 11100101 | 229 |
| 10000110 | 134 | 10100110 | 166 | 11000110 | 198 | 11100110 | 230 |
| 10000111 | 135 | 10100111 | 167 | 11000111 | 199 | 11100111 | 231 |
| 10001000 | 136 | 10101000 | 168 | 11001000 | 200 | 11101000 | 232 |
| 10001001 | 137 | 10101001 | 169 | 11001001 | 201 | 11101001 | 233 |
| 10001010 | 138 | 10101010 | 170 | 11001010 | 202 | 11101010 | 234 |
| 10001011 | 139 | 10101011 | 171 | 11001011 | 203 | 11101011 | 235 |
| 10001100 | 140 | 10101100 | 172 | 11001100 | 204 | 11101100 | 236 |
| 10001101 | 141 | 10101101 | 173 | 11001101 | 205 | 11101101 | 237 |
| 10001110 | 142 | 10101110 | 174 | 11001110 | 206 | 11101110 | 238 |
| 10001111 | 143 | 10101111 | 175 | 11001111 | 207 | 11101111 | 239 |
| 10010000 | 144 | 10110000 | 176 | 11010000 | 208 | 11110000 | 240 |
| 10010001 | 145 | 10110001 | 177 | 11010001 | 209 | 11110001 | 241 |
| 10010010 | 146 | 10110010 | 178 | 11010010 | 210 | 11110010 | 242 |
| 10010011 | 147 | 10110011 | 179 | 11010011 | 211 | 11110011 | 243 |
| 10010100 | 148 | 10110100 | 180 | 11010100 | 212 | 11110100 | 244 |
| 10010101 | 149 | 10110101 | 181 | 11010101 | 213 | 11110101 | 245 |
| 10010110 | 150 | 10110110 | 182 | 11010110 | 214 | 11110110 | 246 |
| 10010111 | 151 | 10110111 | 183 | 11010111 | 215 | 11110111 | 247 |
| 10011000 | 152 | 10111000 | 184 | 11011000 | 216 | 11111000 | 248 |
| 10011001 | 153 | 10111001 | 185 | 11011001 | 217 | 11111001 | 249 |
| 10011010 | 154 | 10111010 | 186 | 11011010 | 218 | 11111010 | 250 |
| 10011011 | 155 | 10111011 | 187 | 11011011 | 219 | 11111011 | 251 |
| 10011100 | 156 | 10111100 | 188 | 11011100 | 220 | 11111100 | 252 |
| 10011101 | 157 | 10111101 | 189 | 11011101 | 221 | 11111101 | 253 |
| 10011110 | 158 | 10111110 | 190 | 11011110 | 222 | 11111110 | 254 |
| 10011111 | 159 | 10111111 | 191 | 11011111 | 223 | 11111111 | 255 |

# APPENDIX B

## ROUTING PROTOCOLS

Routers use a routing protocol to exchange information with one another. Routers utilize this information to build up routing tables that enable them to choose the paths that datagrams will follow to their destinations. A typical routing table entry includes:

- A destination network or destination subnet.
- The IP address of the router that is the next hop on the path to the destination.
- The router interface to be used to reach the next hop router.
- One or more metrics used to evaluate this path to the destination.
- The routing protocol that provided information about this path.
- A time measurement that indicates how fresh this routing information is.

In this appendix, we will briefly sketch the characteristics of routing protocols that might be used between a site router and an ISP router.

## RIP

RIP is a routing protocol that has been around for many years and is popular because it is easy to set up. If a datagram's destination is on a link that is directly connected to a RIP router, the router can package the datagram into a frame and deliver it directly across that link. Otherwise, the RIP router forwards the datagram to the neighboring router that has the shortest hop-count distance to the destination. Each link that is crossed on the way to a destination counts as one hop. The number of hops to a destination is the only metric that RIP uses. However, if a particular link is very slow, it can be counted as the equivalent of several hops.

RIP routers exchange information every 30 seconds. If no report is received from a neighbor router for 3 minutes, then that neighbor is declared to be dead and all routing through the neighbor is canceled. RIP is not suitable for use on dial-up lines. Some organizations have carelessly enabled RIP on ISDN dial-up connections and discovered that their router has kept the line up around the clock so that it could send its RIP reports.

If your ISP wants to exchange RIP information with you, make sure that the ISP does not send you its complete Internet routing table every 30 seconds, because this can choke a slow link pretty thoroughly. A single default entry pointing to the ISP router as the doorway to the Internet usually is the only information that you would need or want.

RIP was upgraded to RIP-2 a few years ago. If you are going to use RIP, it is strongly recommended that you use this current (RIP-2) version.

RIP has some shortcomings:

■ Fifteen is the largest hop count metric that is valid.

■ As a network grows, the routing tables get large. Exchanging them every 30 seconds creates a lot of traffic.

■ It takes 3 minutes to realize that a neighbor is dead. Therefore, it can take a long time to fix bad routes when a router crashes.

■ In version 1 of RIP, a single subnet mask had to be used for all subnets with a given network number. For example, a single mask would have to be used throughout network 130.15. Version 2 lets you use *variable length subnet masks* (VLSM), which means that you can use as many different mask sizes as you need. This is an important feature when an ISP is using RIP to sort out addressing for different sites that are using pieces out of a single network address space.[1]

# Cisco's (E)IGRP

ISPs that use Cisco routers frequently choose to use Cisco's proprietary Interior Gateway Routing Protocol (IGRP) to exchange information with customer sites when this is necessary. The most recent version of IGRP is called Enhanced IGRP or EIGRP.

IGRP routers calculate the cost of getting to a destination using a formula that takes many factors into consideration, including:

■ The delay along a path.

■ The smallest bandwidth among the links along the way.

■ The current load, that is, how much of the bandwidth is currently in use.

■ The reliability, which is the percentage of the datagrams that are being delivered undamaged.

---

[1] Recall that some ISPs give small customers a fragment of a Class C address space.

- The number of hops to the destination.
- The largest Maximum Transmission Unit (MTU) that can cross every link along the path.

IGRP metrics can be whatever size they need to be to report the information accurately. IGRP keeps information about multiple paths to a destination, and can split traffic across these paths.

IGRP routers exchange information every 90 seconds, but an important change triggers a prompt report. Routers using the old version of IGRP exchange complete routing tables. Routers running the enhanced version (EIGRP) only exchange update information, which cuts down significantly on network overhead traffic.

The enhanced version, EIGRP, supports variable-length subnet masks and prevents datagrams from following circular paths. EIGRP has worked well for large, complicated networks. However, you cannot use it in routers that are built by vendors other than Cisco, unless they have licensed the technology.

# OSPF

The Open Shortest Path First routing protocol was designed to scale to networks of any size and complexity. OSPF is an Internet standard, and can be used by any router vendor.

OSPF scales very well using "areas." A small network is made up of a single area. A large network is broken up into several areas. One area must act as a backbone for the network, as shown in Figure B.1. An ISP's network acts

**Figure B.1**
An OSPF backbone area and other areas.

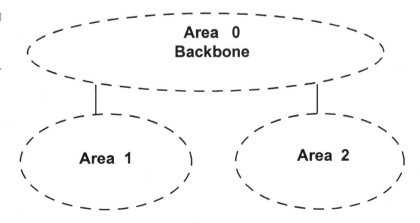

as backbone for its customers' networks, which would correspond to areas. OSPF provides a way to pass information to external networks. An ISP can use this capability to pass summarized routing information to other ISPs.

When the routers within an OSPF area start up, they exchange enough information so that every router actually can build a complete map of the area. Routers also can exchange information about link characteristics such as bandwidth or delay.

- Routing within an area is very accurate, since each router has a complete road map of the area.

- If there are several good paths to a destination, the road map also makes it possible to split traffic across these paths.

- Once the map has been generated, only updates need to be sent. OSPF generates very little overhead traffic.

- The fact that a link has gone down will propagate from a router adjacent to the link to the rest of the routers in the area. All of the routers immediately will adjust their paths.

- Each area reports the networks and subnets that it contains to the backbone. Hence, backbone routers know how to get to anywhere.

- When data has to be delivered to a destination outside its own area, it is routed to the backbone.

OSPF is suitable for large networks, and is attractive to sites that like to use standard protocols.

# Exterior Gateway Protocols

Several *Exterior Gateway Protocols* (EGPs) have been designed specifically to enable ISPs to exchange routing information with one another or with their customers.

## EGP

A very simple exterior protocol that was itself called the *Exterior Gateway Protocol* was designed in the early 1980s. It enables a router to report the list of networks and subnets at its site.

## BGP

Today, most ISPs exchange routing information with one another using the *Border Gateway Protocol* (BGP). Several *traceroute* examples in Chapter 1 demonstrated that Internet traffic often crosses multiple ISP networks. BGP information builds up path data that shows the chain of networks that are crossed on the way to a destination. To be more precise, BGP assembles the chain of *Autonomous Systems* that are traversed. An Autonomous System is made up of one or more networks under a common administration with a consistent routing policy. Every Autonomous System is assigned an identification number.

A very large and very complex site that is connected to more than one Internet Service Provider may have a need to exchange BGP information with its ISPs. In order to do this, the site must obtain an Autonomous System number. There is a registration template at the main registration site (currently the InterNIC Registration Service) that is used for this purpose.

# APPENDIX C

## UNIX CONFIGURATION

When an administrator configures Unix networking for the first time, the process often seems confusing and disorganized. The goal of this appendix is to explain the logic behind the process, and map out where each part of the information usually is located. Many vendors have created versions of the Unix operating system, and sometimes a vendor will depart slightly from these conventions.

Unix was built to be flexible. To carry out this philosophy, each Unix command is designed to do a single, sharply focused task. This allows an administrator to write scripts that combine commands to perform any job that needs to be done.

Unix network configuration is carried out by a combination of:

- Scripts that are executed when the system boots. Start-up scripts may be stored under */etc* or */etc/rc* and have names like *rc.boot, rc.local*. Sometimes scripts are stored in a series of subdirectories under */etc*. These subdirectories usually have names that start with *rc*.

- Files containing parameters that are read at boot time.

- Files containing parameters that are read as needed while the system is running.

There is another important component. A special process called *inetd* runs all the time. It is in control of services that are started up on demand, when a client request for the service arrives. The services that it controls are listed in a configuration file called */etc/inetd.conf*.

## Configuring a Unix TCP/IP Server

TCP/IP is very democratic. The basic information that is needed to add a Unix system to a TCP/IP network is the same as the information needed by a PC or Macintosh:

- IP address
- Subnet mask
- Default router
- Addresses of Domain Name Servers

An Internet server needs to have a stable, permanent IP address. Therefore, a server's configuration information either should be configured manually, or else a permanent IP address should be configured at a network boot server.[1] A server's name and address must be registered in a public DNS database.

If configuration information will be obtained from a network server, then the command or commands that are needed to launch communication with a network boot server are placed into a start-up script. We will focus on the case where configuration is carried out manually at the system itself.

# IP Address and Subnet Mask

When manual configuration is used, a Unix system's IP address and subnet mask are entered via the *ifconfig* command. This command must be repeated every time the system boots, and so it needs to be placed into a start-up script.

The same command can be used to view the current address and mask settings. The example that follows shows settings for two Ethernet interfaces (le0 and le1) and for the dummy loopback interface. The display below presents subnet masks in hexadecimal (ffffff00) instead of dotted decimal (255.255. 255.0).

```
> ifconfig -a
le0: flags=63<UP,BROADCAST,NOTRAILERS,RUNNING>
     inet 128.36.23.1 netmask ffffff00 broadcast 128.36.23.255
le1: flags-63<UP,BROADCAST,NOTRAILERS,RUNNING>
     inet 128.36.0.22 netmask ffffff00 broadcast 128.36.0.0
lo0: flags=49<UP,LOOPBACK,RUNNING>
     inet 127.0.0.1 netmask ff000000
```

# Default Route

A default route is configured manually by means of the *route* command. This command must be repeated every time the system boots and so for manual configuration it needs to be placed into a start-up script. The *route* command can assign a Maximum Segment Size and receive window size for connections

---

[1] Recall that one method of configuring TCP/IP systems is to allow a Dynamic Host Configuration Protocol (DHCP) boot server to provide systems with their addresses, subnet masks, default routers, Domain Name Servers, and other parameters. There are several other automatic configuration methods that have been used for Unix systems over the years.

that use the default route. Recall that these are very important performance parameters.

If a server is located on a LAN that has more than one router, then sometimes it is convenient to set up additional static routes. As many *route* commands as are needed can be entered into a start-up script.

# Domain Name Server

Domain name servers are listed in a file called */etc/resolv.conf*. A sample file is:

```
vnet.net
166.82.1.3
166.82.1.8
```

The *vnet.net* domain name is used to complete local names. For example, suppose that a user types:

```
telnet katie
```
The name *katie* will automatically be completed to *katie.vnet.net*.

# Unix Services

Information about services that will run at the Unix server is spread across several files:

- There is a file (*/etc/services*) that matches service names (such as *ftp* or *domain*) to port numbers.
- Recall that a file (*/etc/inetd.conf*) lists programs that should be started when a client request arrives.
- Other service programs run constantly in the background. These will be invoked from one of the scripts that is run at system startup.

  We'll provide details in the sections that follow.

# Matching Services to Ports

The */etc/services* file matches service applications with the ports at which they will run. Sample entries include:

```
ftp-data    20/tcp
ftp         21/tcp
telnet      23/tcp
smtp        25/tcp    mail
whois       43/tcp    nicname    # usually to sri-nic
domain      53/tcp
domain      53/udp
bootps      67/udp               # bootp server
bootpc      68/udp               # bootp client
tftp        69/udp
pop-3       110/tcp              # PostOffice V.3
```

A file exactly like this also is used at NT servers.

# Starting Services

A service can be started automatically at boot time. This is done by putting a command including the program's pathname and parameters into one of the scripts that are run automatically when the system boots.[2]

Other services are started on demand, when a client request arrives. These services must be listed in the file called */etc/inetd.conf*. A typical entry looks like:

| # service_name | socket_type | proto | flags | user | server_pathname | args |
|---|---|---|---|---|---|---|
| ftp | stream | tcp | nowait | root | /usr/local/etc/in.ftpd | in.ftpd -l |

When a client process connects to TCP port 21, the */etc/services* file is searched to find the matching service name, which in this case is *ftp*. Then *ftp* is looked up in the */etc/inetd.conf* file. The *ftp* entry identifies the full pathname for the service program. The args value repeats the name of the program and includes other parameters (in this case, -l to request logging).

# Location of Service Programs

The location of service programs varies according to the conventions of various Unix subcultures and the preferences of the system administrators. Typical locations are */usr/sbin*, */usr/etc*, or */usr/local/etc*. On some systems, Internet service program names start with "in." ( for example, *in.ftpd, in.telnetd*.

---

[2] Sometimes separate scripts are defined for each service and are stored in a subdirectory. Then a parent script starts all of these subscripts

## Service Programs

Every Unix vendor bundles its own implementations of the standard Internet services with its systems. Many of these are based on Berkeley Unix, so there are strong similarities between the services across platforms. There also are some differences, and so you will need to check the manuals for your own systems.

## Service Configuration Files

Each service has its own implementation-specific configuration files. For example, many Unix vendors provide file transfer server implementations based on the Berkeley server. For these servers, a configuration file called */etc/ftpusers* is used to define server access privileges. But for another popular file transfer server created at Washington University in St. Louis, access is controlled by a file called */etc/ftpaccess*.

In Chapter 12, we described the files used to configure a Domain Name Server. These are quite consistent across different platforms. And in Chapter 14, we described configuration files for the Apache Web server. Each type of Web server has very different configuration files.

# GLOSSARY

**Access Control**  A facility that defines each user's privileges to access computer data.

**Acknowledgment**  TCP requires that data be acknowledged before it can be considered to have been transmitted safely.

**ActiveX**  A unit of object-oriented software (called a "control"), written using Microsoft conventions. An ActiveX control can be downloaded by a Web browser and executed on a Windows desktop. An ActiveX control has full access to the client operating system. For example, it can read and write local files.

**Address Resolution Protocol (ARP)**  A protocol that dynamically discovers the physical address of a system, given its IP address.

**Agent**  Software that is added to a system to perform some specific task automatically.

**Alternate Mark Inversion (AMI)**  A method of encoding 0s and 1s onto telephone wires.

**American National Standards Institute (ANSI)**  Organization responsible for coordinating United States standardization activities.

**Applet**  A program, usually written in Java. Often denotes a small program that is downloaded from a Web server by a browser and executed at the client's computer. Java applets do not have full access to the client operating system and cannot read or write local files.

**Application Programming Interface (API)**  A set of routines that enable a programmer to use computer facilities. The socket programming interface and the Transport Layer Interface are APIs used for TCP/IP programming.

**Application Proxy Firewall**  A firewall that protects users by performing all direct Internet application communications on their behalf.

**ARPANET**  The world's first packet-switching network, which for many years functioned as an Internet backbone.

**ASCII**  American National Standard Code for Information Interchange. Seven of the eight bits in an octet are required to define an ASCII character.

**Asynchronous Transfer Mode**  A switch-based technology that transports information in 53-octet cells. ATM may be used for data, voice, and video.

**Asynchronous Transmission**  Byte-oriented data transmission. Each byte is introduced by a start bit and terminated with one or more stop bits.

**Authentication**   Verification of the identity of a communications partner.

**Authentication Header**   An IP-layer header that authenticates a source and protects the integrity of the data. An Authentication Header normally is inserted after the main IP header and before the other information being authenticated.

**Authoritative DNS Server for a Domain**   A primary or secondary server that includes the information for the domain.

**Authorization**   Setting up privileges to access systems, data, or applications.

**Autonomous System (AS)**   One or more networks under a common administration with a consistent routing policy.

**Backward Explicit Congestion Notification**   For frame relay, a flag that indicates that there is congestion on the backward path.

**Bandwidth**   The quantity of data that can be sent across a link, typically measured in bits per second.

**Baud**   A unit of signaling speed equal to the number of times per second that a signal changes state. If there are exactly two states, the baud rate equals the bit rate.

**Berkeley Software Distribution (BSD)**   UNIX software from the University of California at Berkeley that included TCP/IP support.

**Best Current Practices (BCP)**   A classification applied to a useful RFC that does not define a protocol standard.

**BIND Software**   Domain Name Server software from University of California at Berkeley.

**Biometric**   A measurable physical characteristic that can be used to verify the identity of a person.

**Bipolar 8 Zero Substitution (B8ZS)**   A method of encoding 0s and 1s onto telephone wires that replaces a sequence of 8 0s with a special signal pattern.

**Bootstrap Protocol (BOOTP)**   Old version of a protocol that can be used by booting systems to obtain network configuration information.

**Border Gateway Protocol (BGP)**   A protocol used to exchange Internet routing information. BGP can describe paths that pass through a chain of Internet Service Providers.

**Bounce**   The return of a piece of mail that cannot be delivered.

**Bridge**   A device that connects two or more physical segments of a LAN and forwards frames that have source and destination addresses on different segments.

**Broadcast**   A link frame addressed to all systems on the link.

**Brouter**   A device that performs both bridging and routing functions. Some traffic is selected for routing, while the rest is bridged.

**Browser**   A client application used to access World Wide Web servers and other Internet servers.

**Buffer**   An area of storage used to hold input or output data.

**Caching Only Domain Name Server**   A DNS server that makes queries on behalf of local users (and caches answers), but does not respond to queries from the outside world.

**Canonical Name**   A host's unique true name.

**Certification Authority**   A trusted third party that vouches for the identities of owners of public encryption keys.

**Certificate**   Information created by a trusted third party that validates the identity of a user or server.

**Challenge Handshake**   An authentication scheme that is superior to the conventional method in which a user sends a username and password to a remote system. The scheme is based on a challenge and a response.

**Channel Service Unit (CSU)**   A device that provides a physical interface to the telephone network and performs low-level functions like setting transmit and receive levels.

**Cipher-Block Chaining**   A popular option for DES encryption. A block of already encrypted data is fed into the algorithm as it encrypts the next block.

**Circuit Proxy Firewall**   A firewall that screens systems from the Internet. A user connects to the circuit proxy system, the proxy opens a connection to the desired Internet location, and the proxy copies bytes from the user connection to the Internet connection.

**Classless Inter-Domain Routing (CIDR)**   A method of routing used to enable the network part of IP addresses to consist of a specified number of bits.

**Common Gateway Interface (CGI)**   A standard Application Programming Interface used to pass data to programs that run at a World Wide Web server.

**Communications Protocol**   A set of rules that must be followed by communicating partners.

**Confidentiality**   Protection of information from disclosure to unintended parties.

**Connection**   A logical communication path between TCP users.

**Contact**   In the registration process, a person or role who will be responsible for administering or operating a Domain Name database, or for paying the registration bill.

**Control Connection**   A connection that is used to send File Transfer Protocol requests to a server, and receive status responses from the server.

**COPS**   A package of auditing routines that check the security of a Unix host.

**CRACK**   A password breaking tool used to test Unix password files.

**Cracker**   Someone who attempts to break into computer systems, often with malicious intent.

**Cryptographic Checksum**   A mathematical calculation on the bits in a message or file information. Its purpose is to detect whether information has been altered by repeating the calculation at a later time.

**Cyclic Redundancy Check (CRC)**   The value obtained by applying a mathematical function to the bits in a frame, and appended to the frame. The CRC is recalculated when the frame is received. If the result differs from the appended value, then the frame is discarded. The CRC value is called the frame check sequence.

**D4 Superframe**   An old method (based on 12 frames) used to format digital telephony transmissions.

**Data Circuit-terminating Equipment (DCE)**   Equipment required to connect a DTE to a line or to a network.

**Data Encryption Standard (DES)**   A symmetric encryption protocol officially sanctioned by the U. S. government. There are several options for the manner in which DES is applied. (See Cipher-Block Chaining.)

**Data Exchange Interface (DXI)**   Equipment used to connect to ATM or SMDS services.

**Data Service Unit (DSU)**   A device that interfaces to the telephone network and synchronizes the clocking used to transmit bits with telephone network clocking.

**Data Terminal Equipment (DTE)**   A source or destination for data. Often used to denote terminals or computers attached to a wide area network.

**Datagram**   The unit of information that IP routes across a network.

**DECnet**   Digital Equipment Corporation's proprietary network protocol.

**Demilitarized Zone**   A local area network that acts as a security buffer zone between a private site and the Internet.

**Directory Access Protocol (DAP)**   Client-to-server protocol used to access an X.500 directory service.

**Discard Eligibility**   For frame relay, a flag that indicates which frames should be discarded first if the network becomes congested.

**DIX Ethernet**   A version of Ethernet developed by Digital, Intel, and Xerox.

**Domain Name**  A name made up of one or more labels separated by periods, constructed according to Internet naming conventions.

**Domain Name Server**  A directory server used to translate between computer names and addresses, and to locate Mail Exchangers.

**Domain Name System (DNS)**  A set of distributed databases providing information such as translation between system names and their IP addresses, and the location of mail exchangers.

**DS1**  A frame and interface specification for synchronous T1 lines.

**DS3**  A frame and interface specification for synchronous T3 lines.

**Electronic Signature**  Information enclosed with a message that reliably associates the message with its sender. Currently an encrypted message digest is used for this purpose.

**Encapsulating Security Payload (ESP)**  A protocol designed to provide confidentiality (and optionally, authentication and integrity) to IP datagrams. ESP can be used between a pair of hosts, between a pair of routers, or between a host or router and multiple other hosts and routers.

**Encapsulation**  The action of wrapping a frame header and trailer around a datagram so that the datagram can be transported across a link.

**Encryption**  Transformation of information into a form that cannot be understood without possession of a secret ("decryption") key.

**Enhanced Internet Gateway Routing Protocol (EIGRP)**  The updated version of the proprietary Cisco routing protocol.

**Extended Superframe (ESF)**  Current method (based on 24 frames) used to format digital telephony transmissions.

**Exterior Gateway Protocol (EGP)**  An exterior routing protocol that has been made obsolete by BGP. Routers in neighboring Autonomous Systems used this protocol to identify the set of networks that could be reached within or via each of them.

**Fiber Distributed Data Interface (FDDI)**  A standard for high-speed data transfer across a dual fiber-optic ring.

**File Transfer Protocol (FTP)**  The TCP/IP protocol that enables users to copy files between systems and perform file management functions, such as renaming or deleting files.

**Filtering**  The process of examining incoming and outgoing traffic and discarding items that do not satisfy security rules.

*finger*  A program that displays information about one or more remote users.

**Firewall**   A system that protects a site by controlling the traffic that enters and leaves the site.

**Forward Explicit Congestion Notification (FECN)**   For frame relay, a flag that indicates that there is congestion on the forward path.

**Fragment Offset**   A field in an IP header. If an IP datagram has been fragmented, this field indicates the position of an enclosed fragment, relative to the beginning of the original datagram.

**Fragmentation**   Partitioning of a datagram into pieces. This is done when a datagram is too large for a network technology that must be traversed to reach the destination.

**Frame**   Unit of data sent across a link.

**Frame Check Sequence (FCS)**   A mathematical function applied to the bits in a frame, and appended to the frame. The FCS is recalculated when the frame is received. If the result differs from the appended value, then the frame is discarded.

**Frequently Asked Questions (FAQ)**   A document in the form of questions and answers that summarizes information for a news group or mailing list.

**Fully Qualified Domain Name (FQDN)**   A complete computer name, such as *www.whitehouse.gov*.

**Gateway**   An IP router. Many RFC documents use the term gateway rather than router.

**Gopher**   An obsolete protocol that enables clients to access data at a server by means of a series of menus.

**Graphics Interchange Format (GIF)**   A popular format for graphical image files.

**Guardian**   A Domain Name System contact whose communications are authenticated.

**Handle**   An alphanumeric string that is used as an index that points to a table or database entry.

**High Density Bipolar 3-Zero Maximum Coding (HDB3)**   A method (used in Europe) of encoding 0s and 1s onto telephone wires that replaces a sequence of 4 0s with a special signal pattern.

**High Level Data Link Control Protocol (HDLC)**   A standard that is the basis for several point-to-point link layer protocols.

**High Performance Parallel Interface (HIPPI)**   A high speed communications technology defined by an ANSI standard. Devices communicate via HIPPI across short distances at speeds of 800 and 1600 megabits per second.

**High-Speed Serial Interface (HSSI)**   A serial interface that can support high transmission rates, such as T3.

**Home Page**   The default hypertext page that is sent to a client who has connected to a Web host.

**Host**   In the TCP/IP context, any computer that can execute applications and can communicate via TCP/IP communications protocols.

**Hunt Group**   A set of telephone lines that are accessed via a single telephone number. An incoming call is assigned to the first free line.

**Hypertext**   Text that includes links to files that contain related information.

**Hypertext Markup Language (HTML)**   A set of markup commands used to write hypertext documents. Tags in the document identify elements such as headers, paragraphs, lists, and tables.

**Hypertext Transfer Protocol (HTTP)**   Protocol used between a client and a World Wide Web server.

**Initial Sequence Number (ISN)**   A sequence number defined during TCP connection setup. Data octets sent over the connection will be numbered starting from this point.

**Integrated Services Digital Network (ISDN)**   A telephony technology that provides digital voice and data services.

**Interior Gateway Protocol (IGP)**   Any routing protocol used within a network.

**International Organization for Standardization (ISO)**   An international body founded to promote international trade and cooperative progress in science and technology.

**International Telecommunications Union (ITU)**   A body that oversees several international organizations devoted to communications standards and cooperation.

**International Telecommunications Union Telecommunication Standardization Sector (ITU-T)**   Presides over study groups and writes "Recommendations" for international communications standards.

**internet**   A set of networks connected by IP routers and appearing to its users as a single network.

**Internet**   The world's largest network; the Internet is based on the TCP/IP protocol suite.

**Internet Architecture Board (IAB)**   An Internet Society group responsible for promoting protocol development, selecting protocols for Internet use, and assigning state and status to protocols.

**Internet Assigned Numbers Authority (IANA)**  The authority responsible for controlling the assignment of a variety of parameters, such as IP addresses, well-known ports, multicast addresses, terminal identifiers, and system identifiers.

**Internet Control Message Protocol (ICMP)**  A protocol that is required for implementation with IP. ICMP specifies error messages to be sent when datagrams are discarded. ICMP also provides several useful query services.

**Internet Engineering Steering Group (IESG)**  A group that coordinates the activities of the IETF working groups and performs technical reviews of standards.

**Internet Engineering Task Force (IETF)**  A set of working groups made up of volunteers who develop and implement Internet protocols.

**Internet Gateway Routing Protocol (IGRP)**  A proprietary Cisco routing protocol.

**Internet Group Management Protocol (IGMP)**  A protocol that is part of the IP multicast specification. IGMP is used to carry group membership information.

**Internet Mail Access Protocol (IMAP)**  An electronic mail protocol designed for use with servers that can act as permanent mail repositories.

**Internet Protocol (IP)**  The protocol responsible for transporting datagrams across an internet.

**Internet Server Application Programming Interface (ISAPI)**  A Web server programming interface defined by Microsoft.

**Internet Service Provider (ISP)**  An organization that sells Internet connectivity services.

**Internet Society (ISOC)**  An international organization formed to promote the growth and continued technical enhancement of the Internet.

**InterNIC**  Registration and data services agencies.

**Intranet**  An organization's private TCP/IP network, usually supporting internal information publication via World Wide Web servers.

**Inverse ARP**  A protocol that enables a frame relay interface to discover the IP addresses of interfaces at the remote ends of circuits.

**IP Address**  A 32-bit quantity that identifies a network interface.

**IP Datagram**  The unit of data routed by IP.

**IP Security Option**  In version 4, an optional field in the IP header that contains a security label. The option was designed for use by military and government agencies.

**Java**  A computing language whose code runs in a portable environment.

**Joint Photographic Experts Group (JPEG)**  A specification for an image compression scheme.

**Kerberos**  An authentication service developed at the Massachusetts Institute of Technology. Kerberos uses encryption to prevent intruders from discovering passwords and gaining unauthorized access to files or services.

**Lightweight Directory Access Protocol (LDAP)**  Simplified client-to-server protocol used to access an X.500 directory service.

**Link**  A medium over which systems can communicate using a link layer protocol.

**Link (Web Hyperlink)**  A pointer within a hypertext document that points to another document.

**Listen Queue**  A queue area where incoming TCP session requests are held until the client and server have exchanged their session start-up information.

**Local Area Network (LAN)**  A data network intended to serve an area of only a few square kilometers or less, and consisting of a single subnetwork.

**Local Loop**  The copper wires between a building and a telephone central office.

**Local Management Interface (LMI)**  A set of messages that check that the link to the frame relay service is up and query the status of circuits.

**Logical Link Control (LLC)**  A layer 2 (data link layer) protocol that governs the exchange of data between two systems connected to the same link.

**Loopback Address**  Address 127.0.0.1, used to connect a client to a server running at the same computer.

**MAC Address**  A physical address assigned to a LAN interface.

**MAC Protocol**  A Media Access Control protocol that defines the rules that govern a system's ability to transmit and receive data on a medium.

**Mail Bombing**  Causing inconvenience or denial of service by flooding a recipient with a huge amount of electronic mail.

**Mail Exchanger**  A system used to relay mail into an organization's network.

**Mail Gateway**  A system that performs a protocol translation between different electronic mail delivery protocols.

**Management Information Base (MIB)**  The set of definitions of network-manageable objects. Also, the configuration, status, and performance information that can be retrieved from a network device.

**Maximum Segment Size (MSS)**   The maximum permissible size for the data part of any segment sent on a particular connection.

**Maximum Transmission Unit (MTU)**   The size of the biggest datagram that may be sent across a link.

**Message Digest**   A mathematical calculation on the bits in a message or file. Its purpose is to detect whether information has been altered by repeating the calculation at a later time.

**Message Transfer Agent (MTA)**   An entity that moves electronic mail between computers.

**Multicast IP Address**   An IP address that can be adopted by multiple hosts. Datagrams sent to a multicast IP address will be delivered to all hosts in the group.

**Multihomed Host**   A host that has multiple IP addresses.

**Multipurpose Internet Mail Extensions (MIME)**   Extensions to Internet mail that enable messages to be made of one or more parts. Any part can contain text, image, sound, or application data.

**National Institute of Standards and Technology (NIST)**   A U.S. standards organization that has promoted communications standards.

**Neighbors**   Nodes attached to the same link.

**NetBIOS**   A network programming interface and protocol developed for IBM-compatible personal computers.

**NetBIOS Extended User Interface (NETBEUI)**   A local area network protocol developed for IBM-compatible personal computers.

**Network Access Point (NAP)**   A major switching point at which several Internet Service Providers exchange traffic.

**Network File System (NFS)**   A set of protocols introduced by Sun Microsystems, enabling clients to mount remote directories onto their local file systems, and use remote files as if they were local.

**Network Information Center (NIC)**   An Internet administration facility. A NIC supervises network names and network addresses, and can provide other information services.

**Network Information Service (NIS)**   A set of protocols introduced by Sun Microsystems, used to provide a directory service for network information.

**Network Mask**   A 32-bit quantity with 1s in the locations corresponding to the network part of an address and 0s elsewhere.

**Network Number**   The part of an IP address that identifies a network.

**Non-repudiation** The ability to prove that a source sent specific data, even if the source later tries to deny that fact.

*nslookup* An interactive program used to send queries to a Domain Name Server.

**Open Shortest Path First (OSPF)** A routing protocol that scales well, can route traffic along multiple paths, and uses knowledge of a network's topology to make accurate routing decisions.

**Open Systems Interconnection (OSI)** A set of International Standards Organization standards relating to data communications.

**Opening Handshake** A set of messages exchanged at the start of a TCP session. The messages are used to exchange information about memory resources and maximum segment sizes.

**Packet** Originally, a unit of data sent across a packet-switching network. Currently, the term may refer to a communications protocol data unit for any layer.

**Page** A document unit retrieved from a World Wide Web server.

**Pathname** A character string that identifies a file.

**Payload** The information content carried in a message.

**Perimeter LAN** A local area network that acts as a security buffer zone between a private site and the Internet.

**Physical Address** An address assigned to a network interface.

**Point-to-Point Protocol (PPP)** A protocol for data transfer across serial links. PPP supports authentication, link configuration, and link monitoring capabilities, and allows traffic for several protocols to be multiplexed across the link.

**Point-to-Point Tunneling Protocol (PPTP)** A protocol designed by Microsoft that can be used to authenticate and encrypt traffic between a pair of computers or a computer and a router.

**Port (Device)** A hardware connector in a computer or other device, such as a router.

**Port Number (Application)** A number used to identify an endpoint of a TCP or UDP communication.

**Portable Network Graphics (PNG)** A World Wide Web Consortium standard for platform-independent graphical display.

**Post Office Protocol (POP)** A protocol used to download electronic mail from a server to a client (usually at a desktop system).

**Pretty Good Privacy (PGP)**   An implementation of security software that can be used to authenticate and encrypt data (e.g., transmitted via electronic mail or file transfer).

**Primary Domain Name Server**   A DNS server at which information is entered and updated.

**Private Enterprise Addresses**   Addresses that are not used on the Internet, but have been set aside for use within private networks.

**Protocol Data Unit (PDU)**   A generic term for the protocol unit (e.g., a header and data) used at any layer.

**Protocol Stack**   A family of protocols that work together to provide computer-to-computer communications.

**Provisioning**   Configuration of telephony equipment.

**Proxy Firewall**   A system that acts as an intermediary between users on a private network and Internet servers.

**Public/Private Key Pair**   A pair of keys used for encryption and decryption. A public key can be widely known and used to send its owner encrypted data that is decrypted using a matching private key.

**Receive Window**   The valid range of sequence numbers that a sender may transmit at a given time during the connection.

**Remote Network Monitor (RMON)**   A device that collects information about network traffic.

**Request For Comments (RFC)**   A document describing an Internet protocol or related topics. RFC documents are available online at various Network Information Centers.

**Reseaux IP Europeens (RIPE)**   Coordination center for network registration for Europe.

**Resolver**   Software that enables a client to access the Domain Name System databases.

**Resource Records**   Domain Name System records. The records in a zone file.

**Response Time**   The elapsed time between sending a request and receiving a response.

**Retransmission Timeout**   If a segment is not ACKed within the period defined by the retransmission timeout, then TCP will retransmit the segment.

**Reverse Address Resolution Protocol (RARP)**   A protocol that enables a computer to discover its IP address by broadcasting a request on a network.

**Root Domain Name Database**  A database that ties the entire Domain Name System together. It provides the locations of top-level country Domain Name Servers and many second level Domain Name Servers.

**Round Trip Time (RTT)**  The time elapsed between sending a TCP segment and receiving its ACK.

**Router**  A system that forwards traffic.

**Routing Information Protocol (RIP)**  A simple protocol used to exchange information between routers.

**Routing Metric**  A measurement that enables a router to evaluate the efficiency of a particular path.

**Routing Policy**  The sets of sources and destinations for which an Autonomous System is willing to route traffic.

**Routing Registry**  A database containing route information, used to forward data along a path that traverses two or more Autonomous Systems.

*RWhois*  An application that automatically retrieves *Whois* database information from an appropriate *Whois* database, wherever it may be located.

**Second Level Domain Name**  The combination of the final two labels in a computer name, such as *ibm.com* or *yale.edu*.

**Secondary Domain Name Server**  A DNS server that obtains its database from a primary DNS server.

**Secure Electronic Transaction (SET)**  A set of standards for performing secure electronic transactions using credit cards or bank cards.

**Secure Hypertext Transfer Protocol (S-HTTP)**  A protocol that enables Hypertext Transfer Protocol (World Wide Web) clients and servers to identify themselves reliably and communicate in a way that prevents tampering or eavesdropping.

**Secure Sockets Layer (SSL)**  A protocol that enables clients and servers to identify themselves reliably and communicate in a way that prevents tampering or eavesdropping. SSL can be used to add security to any application.

**Security Association**  A communication protected by a specific selection of security parameters.

**Security Gateway**  A system that provides security to datagrams sent from internal systems to external systems.

**Segment**  A TCP header and optionally, some data.

**Send Window**  The range of sequence numbers between the last octet of data that already has been sent and the right edge of the receive window.

**Sequence Number** A 32-bit field of a TCP header. If the segment contains data, the sequence number is associated with the first octet of the data.

**Serial Line Interface Protocol (SLIP)** A very simple protocol used for transmission of IP datagrams across a serial line.

**Server Message Block (SMB)** A protocol that supports information exchange for NETBIOS/NETBEUI local area networks.

**Service Profile Identifier (SPID)** A number that serves as an index that points to the service choices contracted for an ISDN line.

**Shadow Password** For Unix, the practice of using system software that maintains user passwords in a file that is hidden from ordinary users.

**Shortest Path First** A routing algorithm that uses knowledge of a network's topology in making routing decisions.

**Simple Key-management for Internet Protocols (SKIP)** A protocol that provides key management for IP security.

**Simple Mail Transfer Protocol (SMTP)** A TCP/IP protocol used to transfer mail between systems.

**Simple Network Management Protocol (SNMP)** A protocol that enables a management station to monitor network systems and receive trap (alarm) messages from network systems.

**Slow Start** A procedure that TCP uses at session start-ups and during network congestion. The procedure prevents the network from being overloaded with a burst of new traffic.

**Sniffing** Eavesdropping on a network.

**Socket Address** The full address of a communicating TCP/IP entity, made up of a 32-bit network address and a 16-bit port number.

**SOCKS** A popular circuit proxy firewall specification and implementation.

**Source Quench** An ICMP message sent by a congested system to the sources of its traffic.

**Source Route** A sequence of IP addresses identifying the route a datagram must follow. A source route may optionally be included in an IP datagram header.

**Spoofing** Using a forged IP address.

**Spool** Save to a temporary directory.

**Standard Generalized Markup Language (SGML)** A powerful markup language used to describe elements in portable documents.

**Start Bit** A bit code used to mark the beginning of a byte that is sent via asynchronous transmission.

**Stop Bits**   One or more bit codes used to mark the end of a byte that is sent via asynchronous transmission.

**Subnet Address**   A selected number of bits from the local part of an IP address, used to identify a set of systems connected to a common link.

**Subnet Mask**   A 32-bit quantity, with 1s placed in positions corresponding to the network and subnet parts of an IP address.

**Subnetwork**   An Ethernet LAN, Token-Ring LAN, point-to-point line, frame relay circuit, or any other type of local area network or wide area connection.

**Switched Multimegabit Data Service (SMDS)**   A data transfer service developed by Bellcore.

**SYN**   A segment used at the start of a TCP connection. Each partner sends a SYN containing the starting point for its sequence numbering, and, optionally, the size of the largest segment that it is willing to accept.

**Synchronous Optical Network (SONET)**   A telephony standard for the transmission of information over fiber optic channels.

**Synchronous Transmission**   Bit-oriented data transmission. A timing signal is used to indicate where bits begin and end.

**T1**   A digital telephony service that operates at 1.544 megabits per second. DS1 framing is used.

**T3**   A digital telephony service that operates at 44.746 megabits per second. DS3 framing is used.

**Tag (HTML Tag)**   A bracketed instruction included in a hypertext document that describes how the document should be formatted or describes a link to another document.

*Telnet*   The TCP/IP application protocol that enables a terminal attached to one host to login to other hosts and interact with their applications.

**Throughput**   The amount of data transmitted across a circuit, measured in bits per second.

**Time-to-Live (TTL)**   A limit on the length of time that a datagram can remain within an internet. The Time-to-Live usually is specified as the maximum number of hops that a datagram can traverse before it must be discarded.

**Time To Live (TTL)**   Domain Name System Time To Live. A limit on the length of time that a DNS server can use a record that it has obtained from an authoritative server.

**Tn3270**   *Telnet,* used with options that support IBM 3270 terminal emulation.

**Token-Ring**   A local area network technology based on a ring topology.

**Top Level Domain Name**   The rightmost label in a computer name (for example, *com*, *edu*, or *uk*).

**Transmission Control Protocol (TCP)**   A protocol that provides reliable, connection-oriented data transmission between a pair of applications.

**Transmit Queue Size**   The maximum number of datagrams that can be held waiting to be sent out of an interface.

**Transport Layer Interface (TLI)**   An application programming interface introduced by AT&T that interfaces to multiple communications protocols.

**Trivial File Transfer Protocol (TFTP)**   A very basic protocol used to upload or download files. Typical uses include initializing diskless workstations or downloading software from a controller to a robot.

**Trojan Horse**   A program that appears to do useful work, but also includes secret routines that the perpetrator can use to access the victim's data, or to open up access to the victim's computer.

**Unicast Address**   An address assigned to a single interface.

**Uniform Resource Locator (URL)**   An identifier for an item that can be retrieved by a World Wide Web browser. A URL also may identify a program to be run at a server.

**Uniform Resource Name (URN)**   An identifier for an item that can be retrieved by a World Wide Web browser, which provides a generic name. This may map to several locations from which the item may be retrieved.

**Universal Resource Identifier (URI)**   An identifier for an item that can be retrieved by a World Wide Web browser. The identifier may be a Uniform Resource Locator or a Uniform Resource Name.

**Universal Time Coordinated (UTC)**   Formerly known as Greenwich Mean Time.

**Usenet**   Thousands of bulletin board-like news groups whose information is available on the Internet.

**User Agent (UA)**   An electronic mail application that helps an end user to prepare, save, and send outgoing messages and view, store, and reply to incoming messages.

**User Datagram Protocol**   A simple protocol enabling an application to send individual messages to other applications. Delivery is not guaranteed, and messages need not be delivered in the same order as they were sent.

**Variable-Length Subnet Masks (VLSM)**   The use of several different masks on a network in order to define subnets that can contain different numbers of hosts.

**Virtual Circuit**   A virtual circuit is made up of links that are shared between many users, although each circuit appears to its users as a dedicated end-to-end connection.

**Virus**   A routine that attaches to other legitimate programs, and usually harms local data or program execution.

**W3**   The World Wide Web Consortium.

**Well-known Port**   A TCP or UDP port whose use is standardized by the Internet Assigned Numbers Authority.

**Whois**   A database application used to look up Internet site and contact information.

**Wide Area Network (WAN)**   A network that covers a large geographical area. Typical WAN technologies include point-to-point, X.25, and frame relay.

**Winsock**   A TCP/IP application programming interface, based on the standard socket programming interface and adapted for use at Microsoft Windows systems.

**World Wide Web**   A set of Internet servers that enables clients to access information and applications.

**World Wide Web Consortium**   An international consortium whose purpose is to develop common standards for the evolution of the World Wide Web.

**Worm**   A program that replicates itself at other networked sites.

**X11**   A windowing system invented at MIT.

**X400**   A set of international standards for message transfer.

**X/Open**   A consortium of computer vendors, cooperating to provide a common application environment.

**X-Window System**   A set of protocols developed at MIT that enables a user to interact with applications which may be located at several different computers via a graphical user interface.

**Zone**   A part of the Domain Name System database, corresponding to a contiguous part of the naming tree.

# ABBREVIATIONS AND ACRONYMS

| | |
|---|---|
| AADS | Ameritech Advanced Data Services |
| AAL | ATM Adaptation Layer |
| ACK | An Acknowledgment |
| ADSL | Asymmetric Digital Subscriber Line |
| AF | Address Family |
| AH | Authentication Header |
| AMI | Alternate Mark Inversion |
| ANSI | American National Standards Institute |
| API | Application Programming Interface |
| APNIC | Asian-Pacific Network Information Center |
| ARP | Address Resolution Protocol |
| ARPA | Advanced Research Projects Agency |
| ARPANET | Advanced Research Projects Agency Network |
| AS | Autonomous System |
| ASA | American Standards Association |
| ASCII | American National Standard Code for Information Interchange |
| ATM | Asynchronous Transfer Mode |
| B8ZS | Bipolar 8 Zero Substitution |
| BBN | Bolt, Beranek, and Newman, Incorporated |
| BCP | Best Current Practices |
| BECN | Backward Explicit Congestion Notification (Frame Relay) |
| BGP | Border Gateway Protocol |
| BIND | Berkeley Internet Name Domain |
| BOOTP | Bootstrap Protocol |
| BRI | Basic Rate Interface |
| BSD | Berkeley Software Distribution |
| CBC | Cipher-Block Chaining |
| CERT | Computer Emergency Response Team |

| CGI | Common Gateway Interface |
|---|---|
| CHAP | Challenge Handshake Authentication Protocol |
| CIDR | Classless Inter-Domain Routing |
| COAST | Computer Operations, Audit, and Security Technology |
| COPS | Computer Oracle and Password Protection Program |
| CPU | Central Processing Unit |
| CRC | Cyclic Redundancy Check |
| CSLIP | Compressed SLIP |
| CSMA/CD | Carrier Sense Multiple Access with Collision Detection |
| CSR | Certificate Signing Request |
| CSU | Channel Service Unit |
| DAP | Directory Access Protocol |
| DARPA | Defense Advanced Research Projects Agency |
| DCE | Data Circuit-terminating Equipment |
| DCE | Distributed Computing Environment |
| DDN | Defense Data Network |
| DDN NIC | Defense Data Network Information Center |
| DE | Discard Eligibility (Frame Relay) |
| DES | Data Encryption Standard |
| DFS | Distributed File Service |
| DHCP | Dynamic Host Configuration Protocol |
| DIX | Digital, Intel, and Xerox (Ethernet protocol) |
| DLCI | Data Link Connection Identifier |
| DLL | Dynamic Link Library |
| DMTF | Desktop Management Task Force |
| DNS | Domain Name System |
| DS0 | Digital Signaling level 0 |
| DSU | Data Service Unit |
| DTE | Data Terminal Equipment |
| DUA | Directory User Agent |
| DXI | Data Exchange Interface |
| EBCDIC | Extended Binary-Coded Decimal Interchange Code |

| EGP | Exterior Gateway Protocol |
|---|---|
| EIGRP | Enhanced Internet Gateway Routing Protocol (Cisco proprietary) |
| ESF | Extended Superframe |
| ESMTP | Extended SMTP |
| ESP | Encapsulating Security Payload |
| FAQ | Frequently Asked Questions |
| FCS | Frame Check Sequence |
| FDDI | Fiber Distributed Data Interface |
| FDIC | Federal Deposit Insurance Corporation |
| FECN | Forward Explicit Congestion Notification (Frame Relay) |
| FIN | Final Segment |
| FQDN | Fully Qualified Domain Name |
| FTP | File Transfer Protocol |
| FYI | For Your Information |
| GIF | Graphics Interchange Format |
| GMT | Greenwich Mean Time |
| GUI | Graphical User Interface |
| HDB3 | High Density Bipolar 3-Zero Maximum Coding |
| HDLC | High Level Data Link Control Protocol |
| HDSL | High Bit Rate Digital Subscriber Line |
| HINFO | Host Information |
| HIPPI | High Performance Parallel Interface |
| HSSI | High-Speed Serial Interface |
| HTML | Hypertext Markup Language |
| HTTP | Hypertext Transfer Protocol |
| IAB | Internet Architecture Board |
| IANA | Internet Assigned Numbers Authority |
| IBM | International Business Machines |
| ICMP | Internet Control Message Protocol |
| ID | Identifier |
| IEEE | Institute of Electrical and Electronics Engineers |

| IEN | Internet Engineering Notes |
|---|---|
| IESG | Internet Engineering Steering Group |
| IETF | Internet Engineering Task Force |
| IGMP | Internet Group Management Protocol |
| IGP | Interior Gateway Protocol |
| IGRP | Internet Gateway Routing Protocol (Cisco proprietary) |
| IIS | Internet Information Server (from Microsoft) |
| ILMI | Interim Local Management Interface |
| IMAP | Internet Mail Access Protocol |
| INN | Internet News |
| INTA | International Trademark Association |
| IP | Internet Protocol |
| IPng | IP next generation (version 6) |
| IPSO | IP Security Option |
| IPv6 | IP version 6 |
| IPX | Internetwork Packet eXchange (for NetWare) |
| IRQ | Interrupt Request |
| ISAPI | Internet Server Application Programming Interface |
| ISDN | Integrated Services Digital Network |
| ISI | (University of Southern California) Information Sciences Institute |
| ISN | Initial Sequence Number |
| ISO | International Organization for Standardization |
| ISOC | Internet Society |
| ISP | Internet Service Provider |
| ITU | International Telecommunications Union |
| ITU-T | Telecommunication Standardization Sector of the ITU |
| JPEG | Joint Photographic Experts Group |
| KB | Kilobyte |
| Kbps | Kilobits per second |
| LAN | Local Area Network |
| LAPB | Link Access Procedures Balanced |

| | |
|---|---|
| LAPD | Link Access Procedures on the D-channel |
| LDN | Local Directory Number |
| LLC | Logical Link Control |
| LMI | Local Management Interface |
| MAC | Media Access Control |
| MB | Megabyte |
| Mbps | Megabits per second |
| MD5 | Message Digest 5 ( a specific message digest) |
| MIB | Management Information Base |
| MIME | Multipurpose Internet Mail Extensions |
| MS | Millisecond |
| MSS | Maximum Segment Size |
| MTA | Message Transfer Agent |
| MTU | Maximum Transmission Unit |
| MX | Mail Exchanger |
| NAP | Network Access Point |
| NAT | Network Address Translation |
| NCSA | National Computer Security Association |
| NDIS | Network Device Interface Specification |
| NETBEUI | NetBIOS Extended User Interface |
| NETBIOS | Network Basic Input Output System |
| NFS | Network File System |
| NIC | Network Information Center |
| NIS | Network Information System |
| NIST | National Institute of Standards and Technology |
| NLPID | Network Level Protocol ID |
| NNTP | Network News Transfer Protocol |
| NOC | Network Operations Center |
| NS | Name Server |
| NSA | National Security Agency |
| NSF | National Science Foundation |
| NT1 | Network Termination 1 |

| | |
|---|---|
| NTFS | NT File System |
| NTP | Network Time Protocol |
| ODBC | Open Database Connectivity |
| ODI | Open Device Interface |
| OSI | Open Systems Interconnect |
| OSPF | Open Shortest Path First |
| PAP | Password Authentication Protocol |
| PC | Personal Computer |
| PDU | Protocol Data Unit |
| PGP | Pretty Good Privacy |
| PING | Packet Internet Groper |
| PNG | Portable Network Graphics |
| POP | Point Of Presence |
| POP | Post Office Protocol |
| POTS | Plain Old Telephone Service |
| PPP | Point-to-Point Protocol |
| PPTP | Point-to-Point Tunneling Protocol |
| PTR | Pointer |
| PTT | Postal Telegraph and Telephone |
| QoS | Quality of Service |
| RA | Routing Arbiter |
| RARP | Reverse Address Resolution Protocol |
| RAS | Remote Access Service |
| RFC | Request For Comments |
| RIP | Routing Information Protocol |
| RIPE | Reseaux IP Europeens |
| RMON | Remote Network Monitor |
| ROM | Read Only Memory |
| RPC | Remote Procedure Call |
| RR | Routing Registry |
| RST | Reset |
| RSVP | Reservation Setup Protocol |

| | |
|---|---|
| SET | Secure Electronic Transactions |
| SGML | Standard Generalized Markup Language |
| SHA | Secure Hash Algorithm (a message digest) |
| S-HTTP | Secure Hypertext Transfer Protocol |
| SIP | SMDS Interface Protocol |
| SKIP | Simple Key-management for Internet Protocols |
| SLIP | Serial Line Interface Protocol |
| SMAP | Simple Mail Access Program |
| SMB | Server Message Block |
| SMDS | Switched Multimegabit Data Service |
| S/MIME | Secure Multipurpose Internet Mail Extensions |
| SMTP | Simple Mail Transfer Protocol |
| SNMP | Simple Network Management Protocol |
| SOA | Start of Authority |
| SONET | Synchronous Optical Network |
| SPF | Shortest Path First |
| SPID | Service Profile Identifiers |
| SQL | Structured Query Language |
| SSL | Secure Sockets Layer |
| SYN | Synchronizing Segment |
| TCP | Transmission Control Protocol |
| TELNET | Terminal Networking |
| TFTP | Trivial File Transfer Protocol |
| TLI | Transport Layer Interface |
| TTL | Time-To-Live |
| UA | User Agent |
| UDP | User Datagram Protocol |
| URI | Universal Resource Identifier |
| URL | Uniform Resource Locator |
| URN | Uniform Resource Name |
| UTC | Universal Time Coordinated |
| VCC | Virtual Channel Connection |

| | |
|---|---|
| VCI | Virtual Channel Identifier |
| VDSL | Very High-Bit Rate Digital Subscriber Line |
| VLSM | Variable-Length Subnet Masks |
| VPC | Virtual Path Connection |
| VPI | Virtual Path Identifier |
| W3 | World Wide Web |
| W3C | World Wide Web Consortium |
| WAIS | Wide Area Information Service |
| WAN | Wide Area Network |
| WINS | Windows Internet Name Service |
| WIPO | World Intellectual Property Organization |
| WKS | Well Known Service |
| WWW | World Wide Web |
| WYSIWYG | What You See Is What You Get |

# INDEX

# ABOUT THE AUTHOR

**Dr. Sidnie Feit** is Chief Scientist at The Standish Group, a leading Online Transaction Processing and Internet Commerce consulting group. Dr. Feit is the author of *TCP/IP: Architecture, Protocols, and Implementation* and *SNMP: A Guide to Network Management*.